The science and design of the hybrid rocket engine

Richard M. Newlands

Copyright

All of this work is copyright© 2017 Richard M. Newlands apart from illustrations credited at the end of the book.

All rights reserved: no part of this publication may be reproduced, stored in a retrieval system or transmitted in any form or by any means, electronic, mechanical, photocopying, recording, scanning, or otherwise, except as permitted under Sections 107 or 108 of the 1976 United States Copyright Act, without the prior permission of the copyright owner.

ISBN: 978-0-244-60052-5

Dedications

For all the rocketeers who taught me rocketry, especially the late John Stewart, John Bonsor, and John Pitfield. And for all the gentlemen at Southampton University who taught me rocket science.

Thanks to my friends within our rocketry society Aspirespace, the United Kingdom Rocketry Association, and the British Interplanetary Society, who I bounced ideas off. Particular thanks to Dr. Martin Heywood who reviewed the book and provided many useful comments, and Dave Edgar who checked my thermophysics.

Thanks to Jim M. for his rocket testing knowledge and advice.

Thanks to Helen for her support, and book writing and editorial experience.

About the author

Rick 'the Rocketeer' Newlands has been interested in Space travel since childhood.

He graduated with an Honours degree in Aeronautics and Astronautics from Southampton University.

He is currently chairman of the Aspirespace rocket engineering society (www.aspirespace.org.uk), a member of the Scottish Aeronautics and Rocketry Association, a member of the British Interplanetary Society technical committee (www.bis-space.com), and a founder member and council member of the United Kingdom Rocketry Association (www.ukra.org.uk).

He's also a technical advisor for several Space projects.

"The way to discover whether you know something is to try to explain it to someone else."

Disclaimer

The author has strived to maintain the accuracy of information presented herein, but neither the publisher nor the author accept any responsibility for accident or injury arising from following the information given within this book.

Table of Contents

The science and design of the hybrid rocket engine i
Richard M. Newlands i
Copyright ii
Dedications iii
About the author iv
Disclaimer iv
   Table of Contents v
   Introduction viii
      The right toolkit viii
   Chapter 1: A recap of science and maths 1
      Calculus 1
      Forces 3
      Weight and mass 3
      The rocket principle 4
      Newtons's other laws 5
      The kinetic theory of gasses 6
      Temperature 7
      The mole 8
      The Ideal gas law: the trinity of pressure, temperature, and density 8
      Mass continuity and mass flux 9
      Energy 10
      Mixture 17
      Types of rocket engine 17
      The atmosphere 22
      A suborbital trajectory 22
      Indice algebra 23
   Chapter 2: Propellant choices 24
      Fuel selection 24
      The plastics 24
      Metal doping 26
      The oxidisers 26
      Nitrous oxide 27
      Aspects of Nitrous 27
      Subcriticality and Supercriticality 28
      Nitrous properties 28
      Nitrous slosh 33
   Chapter 3: Using Nitrous oxide in rocketry 34
      Filling the run-tank 34
      Vents 35
      Changes in liquid/vapour proportion due to temperature alone 35
      The vapour-only phase 39
      Modelling the Nitrous run-tank emptying 40
   Chapter 4: How a rocket engine works - pressure and flow 44
      Pressure 44

    Mass flow rate and exhaust velocity ... 46
    The thrust equation ... 47
    Nozzle mass flow rate ... 49
    Exhaust velocity and the nozzle shape ... 55
    More pressure ... 60
    In summary ... 60

Chapter 5: How a rocket engine works - performance and heating ... 61
    Nozzle exit velocity and specific enthalpy h ... 61
    Pressure thrust and effective exhaust velocity ... 62
    Exhaust velocity performance ... 65
    Heating the gas ... 67

Chapter 6: Energy and exhaust velocity ... 74
    Chemical energy ... 74
    Combustion ... 76
    The enthalpy of the propellants ... 78
    Thermochemistry ... 79
    Chemical equilibrium ... 80
    Propellant selection ... 83
    PROPEP results: exhaust velocity ... 84
    Isentropic flow ... 85
    Energy losses ... 85
    The nozzle ... 86

Chapter 7: The hybrid combustion chamber ... 88
    The Boundary Layer ... 88
    Regression rate ... 89
    Fuel selection ... 92
    Scaling up a hybrid ... 93
    Ignition systems ... 94

Chapter 8: Nitrous hybrid safety ... 96
    Hazards analysis ... 96
    Pressure vessel and plumbing mechanical safety ... 97
    Hydraulic overpressure, the head space ... 100
    The engine plumbing ... 100
    The combustion chamber ... 101
    Nitrous leaks ... 101
    Terminal flatulence ... 101
    Nitrous oxide decomposition hazard ... 102
    Degreasing procedures ... 105
    Nitrous material compatibility ... 106
    Recommendations ... 107

Chapter 9: Design the engine ... 109
    How big? ... 109
    Initial estimate ... 109
    Numerical integration ... 112
    Programming in Microsoft Office's Excel ... 119
    The sim results ... 124
    The Nitrous run-tank ... 125
    The injector ... 125

    The combustion chamber 127
    The nozzle throat 132
Chapter 10: Build the hybrid rocket engine **133**
    The nozzle 133
    The combustion chambers 133
    The Nitrous run-tank 137
    The injector plates 137
    To stop the rocket melting 138
    To stop the rocket leaking 139
    Tolerance 140
Chapter 11: Testing a hybrid rocket engine safely **141**
    Keep safe 141
    The test stand 141
    Test-site safety 144
    Static-firing misfortunes 144
    On with the rocketry 148
Chapter 12: Flying a hybrid **149**
    Air law 149
    Range safety 149
    Flight weight 149
    Remote operation - the launch system 150
    Alignment 151
    Aerodynamics 152
    Trajectory 152
Appendix 1: Earth's Atmosphere **155**
Appendix 2: Design values for a Nitrous oxide and high-density polyethylene hybrid engine **156**
Appendix 3: Mathematical equations used in rocketry **158**
Appendix 4: Altering the PROPEP 3 input file **169**
Appendix 5: Converting regression data **171**
Appendix 6: Plotting a bell nozzle contour **172**
References **177**
Index **179**
Picture credits **184**

Introduction

"Man must rise above the Earth - to the top of the atmosphere and beyond - for only in that way will he fully understand the world in which he lives."- Socrates.

This book will concentrate on hybrid rocket engines utilising Nitrous oxide as their oxidiser, though the same principles apply to any oxidiser. I will discuss both small hobbyist 'HPR'-sized hybrids, as well as one capable of launching one person into Space.

First off, although many refer to the whole craft as a rocket, it's actually just the engine that is the rocket (some types of rocket engines are also known as rocket motors). The rest of the craft can be called the rocket vehicle.

The right toolkit

If you want to plumb in a new kitchen sink, you'll need the right tools for the job. You'll need spanners (wrenches) to screw some of the plumbing joints together, and a blowtorch to braze the remaining joints together. If you don't have the correct tools, you'll never be able to do the job.

For rocketry, the correct and necessary tools are a vivid imagination to be able to picture in your mind what you need to build before you build it, and most essentially: mathematics, science, and engineering. (And plumbing!)

Chapter 1: A recap of science and maths

Rocket science, as the name suggests, involves many of the sciences: physics, chemistry, thermodynamics, mechanics, and stress analysis, (and also plumbing!). Also required is mathematics: primarily algebra.

So to begin, let's review the core science and mathematics required to design a hybrid engine, some of which you may already know.

Calculus

Isaac Newton, (1642-1727)

You may wonder why we're starting with one of the hardest concepts in mathematics that you learnt in high school. Actually, the core ideas of calculus are simple; your main problem at school - apart from the confusingly arcane language used to describe it - was that nobody bothered to tell you what calculus was *for*.

Isaac Newton co-invented a lot of the methods of calculus because he needed a tool to analyse and predict the motion of falling objects and orbiting planets. That's calculus's main use in rocketry: analysing the rocket vehicle's trajectory as affected by the gravity of the Earth, as we'll do in chapter 9.

Differentiation

Let's start with a high school physics example, the equation:

$$a = \frac{V-U}{t} \quad \text{equ. 1.1}$$

which allows us to calculate the acceleration a of an object from the change in its velocity ($V - U$) over some measured time interval t.

We can also write this as:

$$a = \frac{\Delta V}{\Delta t} \quad \text{equ. 1.2}$$

where the Greek letter Δ (delta) means 'a finite change in'. (ΔV occurs a lot in rocketry.)

Though nobody bothered to tell you in school, these equations are actually specialised cases of **differential calculus**, the specialisation being that *the acceleration is constant over the measured time interval,* i.e. the velocity ramped up or down linearly with time (the velocity was a straight rising or falling line on a velocity-versus-time graph).

If the acceleration *wasn't* constant due to the velocity changing in a complex way with time, then we might want to know the *instantaneous* acceleration of our rocket vehicle at some particular time. We can get this (extremely closely) by using a vanishingly small time interval for our Δt, (but not quite zero).

This is written: instantaneous $a = \frac{\Delta V}{\Delta t}$ 'as Δt tends towards zero' $= \frac{dV}{dt}$

The 'd' signifies '*very* small change in'.

so:

$$a = \frac{dV}{dt} \quad \text{equ. 1.3}$$

which can also be written as:

$$a = \frac{d}{dt}(V) \quad \text{equ. 1.4}$$

This is the general form used in physics, and is described as 'The time rate of change of velocity is acceleration' where 'time rate of change' means in plain English: 'divided by the infinitesimally small change in time'.

The mathematician Leibniz invented the notation $\frac{d}{dt}$, however, Isaac Newton used a different symbology which he called **fluxion notation** to denote the same thing: he'd put a dot above the variable, so he'd write: $a = \dot{V}$

In rocketry, both notations are used.

Integration

Newton also needed to perform the reverse process, starting with acceleration and calculating the resulting change in velocity that would occur. [And so, apparently, did the ancient Babylonians, who used a similar method of calculus to predict the motions of the planets.]

Again, high school physics gave us specialised equations that again assumed that the acceleration was constant over the time interval:

$$V - U = a\,t \quad \text{equ. 1.5}$$

or:

$$\Delta V = a\,\Delta t \quad \text{equ. 1.6}$$

Integration, also called **integral calculus**, extends this special case to deal with acceleration *that changes over time*, which is what generally happens in rocketry as the thrust and mass of the vehicle change with time which changes the vehicle's acceleration with time.

We write: 'Velocity is the time integral of acceleration':

$$\Delta V = \int_{start\ time}^{end\ time} a\,dt \quad \text{equ. 1.7}$$

The curly $'\int'$ shape denotes integration.

We integrate from the start time of our finite time interval (perhaps engine ignition) to whatever is the end time of our time interval (perhaps engine burn-out). These time limits on the integration region are shown at the top and bottom of the $'\int'$ symbol.

We'll add together a whole chunk of little instantaneous accelerations 'a' over the finite time interval of interest, here shown as a series of narrow coloured rectangles of width Δt and height 'a':

(We use a rectangle with an assumed *constant* acceleration across the width of the rectangle. The rectangle has area: height a times width Δt which from equation 1.6 = ΔV).

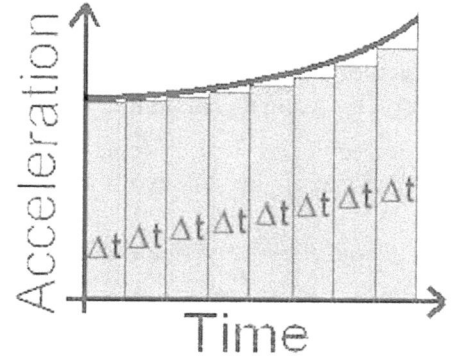

It turns out that the answer we're aiming for (the change in velocity of our rocket vehicle that occurs at the end of our specified time interval) is numerically equal to the area of all the little rectangles shown here added together. This is known as 'the area under the curve' as it is the area under the thick black acceleration curve shown here:

As you can imagine, the smaller we make each time interval Δ*t*, the narrower each rectangle will get (and so we'll need ever more rectangles) but the closer we'll get to the actual area under the thick black curve. The steppy 'roof' made of the tops of all the little rectangles will fit closer to the black curve the narrower the rectangles get.

In the theoretical limit of an infinite number of infinitely narrow rectangles, the area will match exactly the right answer. Remember that a vanishingly narrow (but not quite zero) Δ*t* is written '*dt*' hence the *dt* to the right of the curly integration symbol above.

Integration can be performed algebraically, but it's easier to let a computer do it, as we do in chapter 9. There we'll see that the time interval *dt doesn't* need to be vanishingly small to get a reasonably accurate answer.

Forces

A **force** is a push or pull; generally in rocketry, one that tries to move something.

Forces are measured in Newtons, named after Isaac Newton. (If anyone tries to tell you that Thrust force is measured in kilograms, they're *very* wrong! I'm alarmed that British T.V. science presenters describe forces in kilos.)

When you describe a force, you must always indicate in what direction the force pushes. The direction is very important, because if I push you backwards instead of pushing you forwards, you'll move in a totally different direction. So always conider what direction the force is pushing (or pulling).

To show the direction of a force we often draw the force as an arrow, which is called a force **vector**. The shaft of the arrow shows the direction, and this direction is called the **line of action** of the force.

[A vector is any quantity where direction is as important as its magnitude.]

Forces add vectorially (force arrows are added nose-to-tail).

Weight and mass

A rocket vehicle has weight, and it also has **mass**. These are related but are *not* the same. Mass indicates the number of atoms something contains, which won't change unless you cut it, and is measured in kilograms. 5 kilos of spacecraft remains the same whether they're at ground-level, or out in Space. Gravity doesn't change mass.

Weight and gravity

Earth's gravity pulls every atom (so that's everything with mass) down to the ground. This pull that holds things firmly on the ground is the force we call weight.

$$weight = mass \times gravity \quad \text{equ. 1.8}$$

On the surface of our planet Earth, gravity has a strength of around 9.81 metres per second per second.

Isaac Newton devised a law to describe the effect of gravity: he realised that the gravity of the Earth (or any other planet or star) varies with an inverse square relationship with altitude.

$$g = \frac{\mu}{(R_E+h)^2} \quad \text{equ. 1.9} \quad \text{in metres/second}^2$$

where, for the Earth: μ = 398,600,500 km³s² is the Earth's gravitational constant, R_E is the mean radius of the Earth (6378.14 kilometres), and *h* is the height you're at above the Earth's surface *in kilometres*.

Although somewhat irrelevant in everyday life on the ground, in Space, the difference between mass and weight is very important.

Specific quantities

Quite often, it's easier to deal with quantities per kilogram, such as energy per kilogram. Per-kilogram quantities are known as **mass-specific** quantities, shortened to 'specific'. They're usually denoted by lower-case letters, so if energy is E then specific energy is e.

The rocket principle

It was realised, by Isaac Newton and others, that forces never occur on their own, they always come in pairs. If you create a force, then another force will automatically appear.

Newton referred to the force that you make, the **action**, and he called the force that then automatically occurs, the **reaction**.

Newton also discovered that the reaction has a strength exactly equal to the action, and furthermore, it pushes in exactly the opposite direction to the action.

This was all encapsulated in his 3rd Law of motion, sometimes referred to as the rocket principle.

How it works

The rocket consists of a **combustion chamber**, a large open bottle full of very high pressure gas. You usually heat up the gas in a rocket by burning something, which is termed **combustion**, hence the name.

The hole at the end of the bottle is termed the nozzle; the narrowest part of which is the **nozzle throat**.

If one were to project the circle of the throat onto the inner forward wall of the combustion chamber, we could call this circle the **action area** because the high internal gas pressure pushing on this area generates the majority of the rocket's thrust force.

The internal gas pressure is, of course, pushing forcefully all over the inner walls of the bottle, but everywhere except the action area it cancels out.

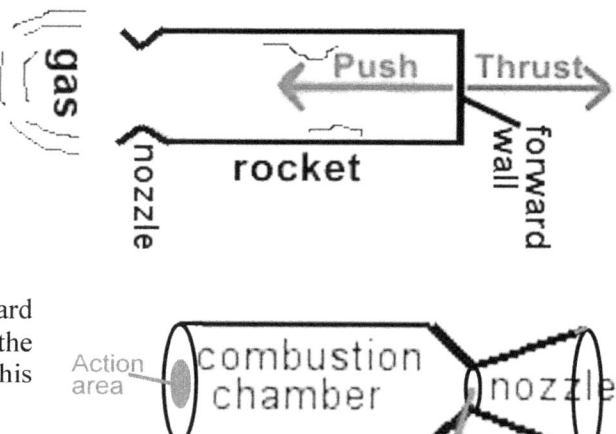

For example, with the rocket lying horizontally as shown here, the pressure is pushing equally on the upper wall as on the lower wall, so no net upper-lower force is generated. All that happens is that the walls get stretched. But opposite the action area there is just a hole, so a net thrust force prevails. The downside (which we could term the 'reaction') is that the precious high pressure gas squirts out the hole so has to be continually replenished. And yet as we'll see shortly, the nature of this outflow is very important for maintaining the internal gas pressure.

The rocket designer tailors the size of the throat, and hence the action area, to achieve the amount of thrust required for the Space mission: <u>thrust force equals gas pressure multiplied by action area.</u> Varying the size of the hole doesn't affect the engine's efficiency.

On the end of the combustion chamber there resides the nozzle, of which the throat is the narrowest part. Its function will be described in chapter 4.

Newtons's other laws

Newton's 1st law: "An object at rest will remain at rest unless acted on by an unbalanced force. An object in motion continues in motion with the same speed and in the same direction unless acted upon by an unbalanced force". ['Unbalanced' here means that if several forces are acting on an object, their combined effect - the resultant force - is not zero.]

This law is often called "the Law of inertia" because **inertia** is the property associated with mass that resists acceleration. Kick a three-tonne satellite in Space and it'll hurt because it resists your attempt to accelerate it.

Newton's 2nd law: Newton originally expressed this law as "The time rate of change of momentum of an object equals the force applied to it divided by its mass." This derivation is of more use to rocketeers (as explained in chapter 4).

Or descriptively, the thrust force F (in Newtons) equals the differentiation with time t (seconds) of mass m (kilograms) times velocity V (metres per second), where the quantity m times V is called **momentum**.

$$F = \frac{d}{dt}(mV) \quad \text{equ. 1.10}$$

Now in the case of an unchanging mass, then m can be taken out of the brackets, and the equation collapses into the familiar high school physics form:

$$F = \frac{d}{dt}(mV) = m\frac{d}{dt}(V) = m\left(\frac{dV}{dt}\right) = ma \quad \text{equ. 1.11}$$

Or "Force equals mass times acceleration" because as we saw earlier on page 1, the rate of change with time t of velocity V is simply acceleration: $a = \dfrac{dV}{dt}$

However, with a rocket, the velocity of the gas mass exiting the rocket nozzle is more or less constant as we'll see in chapter 4, but as this gas mass is continually being lost by the engine out the nozzle, then the remaining propellant mass m (and hence the mass of the entire rocket vehicle) is changing with time, so:

$$F = \frac{d}{dt}(mV) = V\frac{d}{dt}(m) = V\left(\frac{dm}{dt}\right) \quad \text{equ. 1.12}$$

i.e. velocity times the rate that the mass is changing with time in kg/second.

We explore this equation more fully in chapter 4, where we use Newton's fluxion notation (see page 1): $F = V\dot{m}$

[Note that the **Law of conservation of momentum** says that momentum is conserved: it can be redistributed, but its overall value won't change.]

Gas inertia and thrust

Interestingly, Newton's 1st and 2nd laws tell us that if the high-pressure gas within the combustion chamber had no inertia, it would squirt itself out of the chamber instantly. With zero gas residence time within the chamber there would be no high pressure in the chamber. It's the gas's inertia and the ensuing finite rate of increase of gas momentum down the nozzle that sustains the pressure difference between inside the chamber and outside: the gas's flow down the nozzle, and its decreasing pressure drop down the nozzle, support each other in a reciprocal arrangement. No gas inertia would mean no pressure difference and hence no thrust, so the gas inertia is a vital part of the functioning of a rocket.

The kinetic theory of gasses

Daniel Bernoulli developed the first convincing theory of how gasses exert a pressure. He proposed that a gas consists of a large number of particles (which we now know as atoms) which cause pressure by bombarding the walls of their container.

Generally, gas atoms stick together in pairs. For example, it's unusual to find single atoms of oxygen (O) they generally come in twos, which is written O_2.

A clump of joined atoms is called a **molecule**, so O_2 is an oxygen molecule. However, molecules bounce off the walls just as single atoms do, so molecules still cause pressure in the same way.

A gas's molecules are free to move about at high speed. All the tiny molecules bounce off one another like rubber balls. This is called the **kinetic theory** of gas. (Kine is an ancient Greek word for movement.)

If the gas is inside a container, such as a box, the gas molecules also impact the inside walls.

When a molecule hits the wall, it may not change speed, but it does change direction (it rebounds) so its velocity changes, because velocity is a vector (direction is important). Therefore its momentum (its mass times its velocity) changes. By Newton's 3rd law, this momentum change exerts a force on the wall.

Each bounce off a wall creates a really tiny force on the wall. The billions of bounces from the billions of tiny molecules inside add up to what feels like a large steady force, but spread over all of the inside wall. When a force is spread all over a wall like this, we call it a **pressure**.

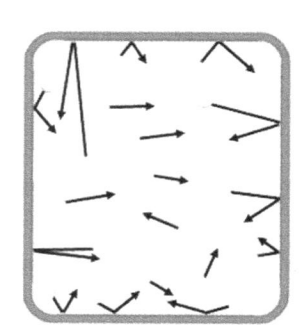

To develop an understanding of gas pressure, science proposed several simplifying assumptions, which although artificial do not detract from the overall picture of what a gas is; the ensuing theoretical model is useful.

- The molecules are treated as structureless particles (are point-like object of 'zero' size).
- The molecules exert no force on each other unless they collide.
- Newton's laws of motion govern the dynamics of the moving molecules, but gravity has a negligible effect.
- All molecular collisions are elastic, which means that no energy is lost in the collisions.
- The molecules are in ceaseless random motion.

It may surprise you that extremely tiny molecules can exert the enormous pressure of the atmosphere (see later) upon you, however the atoms are moving far faster than any bullet: at hundreds or thousands of metres per second. Also, there are an awful lot of them: around 10 to the power 25 (25 zeros) air molecules strike your hand every second.

Components

Let's look more closely at a gas molecule bouncing off one of the inner walls of the combustion chamber. Remember that pressure is the ceaseless bombardment of the walls by millions of molecules every second.

Typically, the molecule hits the wall at an angle, it follows the line drawn in black in the diagram below then hits the wall (on the left in the diagram). Because it's flying along at an upwards angle, it's moving to the left but also upwards, both at the same time.

By drawing a right-angled triangle (called a **vector diagram**) we can show the part of the molecule's movement that acts only to the left (line number 1) and also the part of its movement that goes only upwards (line number 2).

We've **resolved** the molecule's movement into two **components**. A component is that fraction of the whole that acts in the particular direction that you're interested in.

Notice that the molecule's movement upwards (line 2) just slides along the wall, it can't cause a bounce. It's only the molecule's movement to the left (line 1), hitting the wall straight-on, that can cause the molecule to bounce off the wall.

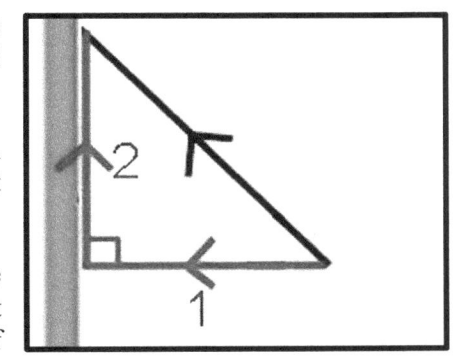

So the force that the molecule's bounce makes on the wall only depends on that part of its motion that hits the wall straight-on (called **normal** to the wall). And so, *pressure always pushes normal to a wall*. When a pressure P acts upon a wall of area A then the resulting force on the wall is equal to the pressure times the wall area A that the pressure acts upon:

$$F = PA \quad \text{equ. 1.13}$$

Pressure within a fluid

But that's when the molecules hit a wall. When the molecules are still far from a wall, they bounce in all directions off the molecules around them, *so the pressure within the gas (not at a wall) pushes in all directions at once*.

But what exactly is the meaning of pressure within a gas? When pressure acts on a wall, that's easy to understand in terms of molecules colliding with the wall. But what is the pressure at some point within the gas? What we really have to understand is that we're talking about a 'force-field' of pressure acting in all radially inward directions on a sphere of gas of very small - but not zero - size. Wide enough to contain perhaps 100 molecules.

But what we really mean is not that there's some fictitious force pushing in on the sphere of gas. Actually, it's all to do with collisions: there's a momentum exchange between the particles outside our imaginary spherical boundary, and those inside. So really they're internal stresses rather than forces at the boundary.

Mixtures of gasses

Mixtures of gasses such as air behave just like a gas of one species. So we can treat the gas within the combustion chamber - which is a mixture of many different gasses - just as if it were a single gas, and refer to it as 'the combustion chamber gas'.

Temperature

Most chemical rockets burn propellants to achieve hot combustion chamber gas.

Heat causes the gas molecules inside the combustion chamber to vibrate, rotate, and fly around at high speed. The more heat you put in, the faster the molecules move in these three ways. The average speed that the molecules move is called **temperature**. Things feel hot or cold depending on how fast the molecules of whatever you're touching are moving.

Kelvin temperature

Before we talk about the effects of temperature, one very important lesson to learn is that in rocketry you can't use degrees C (degrees Celsius or Centigrade) to measure temperature anymore. (Americans can't use Fahrenheit either.) This is because if you use degrees C, any equations that use gas density or gas pressure fail to work.

The problem is where the scientist Anders Celsius (degrees Celsius are also called degrees Centigrade) decided to put zero when he invented his temperature scale. He set zero degrees C as the temperature that water-ice melts.

But you've just learned that temperature is actually the speed at which molecules vibrate and move around. At zero degrees C they're most certainly *not* moving at zero speed, they're moving quite vigorously. For example, oxygen is already a gas at zero degrees C as its vaporizing temperature (when it transforms from a liquid to a gas) is far below zero degrees C. At zero degrees C the oxygen molecules are zipping around at very high speed and create a very high pressure inside a closed tank.

The scientist William Thompson (Lord Kelvin) realised that what was needed was a different temperature scale where zero really meant zero: that is, where the molecules stop moving (zero speed) so they make zero pressure. This zero, called **absolute zero**, is seriously frigid. It's 273.15 degrees *below* zero degrees C.

Kelvin's temperature scale is measured (unsurprisingly) in **Kelvin** (K).

A Kelvin is exactly the same size as a degree C. The only difference is where the zero is, zero Kelvin is set at absolute zero. To convert degrees C to K simply add 273.15

So remember, in rocketry you must use Kelvin (K).

And because Kelvin are always above zero, then it's technically wrong to talk about cold. A cold day is actually nearly 300 Kelvin, the air molecules are zipping about at very high speed. So there's no such thing as cold with Kelvin. There's only varying degrees of hot.

[In old rocketry books or in America you might come across the **Rankine** (R). This has the same zero at absolute zero, but an R is a different size to a K: one R = 5/9ths K i.e. the same size as a Fahrenheit.]

The mole

One measure in chemistry that causes endless confusion is the mole.

A mole is simply one gram of an element (or a substance) times its molecular mass, so is correctly termed a grammole. Rather oddly, it transpires that one gram-mole of a substance contains the same number of atoms as one grammole of any other substance: the exact number of atoms is **Avogadro's number**, of which the latest estimate is $6.02214086 \times 10^{23}$

For example, the **molecular mass** of carbon twelve is 12 (the nucleus of its atom contains six protons plus six neutrons equals 12 constituents) therefore, one gram-mole of carbon twelve has a mass of 12 grams.

Molecular mass (also known as relative molecular mass), is also known somewhat incorrectly as 'molecular weight' which is why it generally has the symbol 'W'.

Confusion occurs because rocketeers are just as likely to use the kilogram-mole (or **kilomole**) rather than the grammole; this is obviously 1,000 times larger. Take care to check which mole unit is being used.

The Ideal gas law: the trinity of pressure, temperature, and density

It's been found by experimentation (and by developing the kinetic theory of gasses) that many gasses at low density and high temperature obey the **perfect gas law** or **Ideal gas law**. This law is a simple equation that says:

$P = \rho R T$ equ. 1.14

P is the pressure of the gas (in Newtons per square metre, also known as **Pascals**) ρ is its density (equal to its mass divided by its volume, in kilograms per cubic metre), and T is its temperature *in degrees Kelvin*. R is just a fixed number called the **specific gas constant** (Joules per <u>kilomole</u> Kelvin) which is different for different gasses.

The Ideal gas law confirms how gasses behave. For example, if you increase the temperature of a gas but keep its pressure constant, then its density drops.

The ideal gas equation can be expanded to include gas mass m and gas volume V:

$$P = \rho R T = \frac{m}{V} R T \quad \text{equ. 1. 15}$$

Gasses at low temperature (or high density) such as Nitrous oxide vapour, deviate from ideal behaviour. They can be described by introducing a fiddle-factor known as the **compressibility factor** Z:

$$P = Z \rho R T \quad \text{equ. 1. 16}$$

Note that R can be derived from the **Universal gas constant** R_0 which has the same value for all gasses, and the particular gas's molecular mass W:

$$R = \frac{R_0}{W} \quad \text{equ. 1. 17}$$

where R_0 has the value 8.3144598(48) kiloJoules per kilomole Kelvin. [Or Joules per mole Kelvin.]

R for air = 0.2869 kJ/kg K

Mass continuity and mass flux

What fluid mass you get out of the end of a pipe is exactly equal to what you put into it, as mass is neither created nor destroyed within the pipe, so there is a continuity of mass between pipe inlet and outlet.

The Law of mass continuity applied to the flow through the pipe can be described by the **mass continuity equation**. If we measure the flow of fluid mass passing any particular point along the pipe which has cross-sectional area A, then:

$$\dot{m} = \rho A V \quad \text{equ. 1. 18}$$

where $\dot{m}$ is Newton's fluxion symbolism for $\frac{dm}{dt}$, the time rate of change of mass, known as the **mass flow rate** (in kilograms per second). ρ and V are the fluid density (kilograms per cubic metre) and velocity (metres per second) at that point in the pipe.

The thrust of a rocket is primarily determined by the mass flow rate of gas passing through the nozzle as we'll see in chapter 4.

Both gasses and liquids are classed as fluids.

Mass flux

Mass flux is equal to the mass flow rate divided by the cross-sectional area of the pipe/port it is being forced to flow through, i.e.

$$mass\ flux = \frac{\dot{m}}{A} = \frac{\rho A V}{A} = \rho V \quad \text{equ. 1. 19}$$

Strictly mass flux is a mass flux *rate*, but is seldom called this. Similarly, mass flux is given the symbol G rather than $\dot{G}$ which it strictly it ought to be. The mass flux of the flow down the port of a hybrid fuel grain determines the speed of the hybrid combustion as we'll see in chapter 7.

[Port is the arcane solid-propellant-motor term for the central hole down the middle of a tubular fuel grain.]

Energy

What is energy? Originally, it was an idea completely made-up by engineers. However, Einstein later declared that mass and energy are the effectively the same thing, and we acknowledge that mass appears to be physically real.

You've probably got a rough idea what energy is; it's a sort of nebulous 'power' that can flow out of one thing and into another, and can change from one form into another. In an electric cooker, electrical energy turns into heat energy. In a car, chemical energy in the petrol (gasoline) turns into energy of movement of the car.

Scientists work with energy because energy makes the mathematical analysis a lot easier: many of the equations listed in appendix 3 were put together using the equations for energy. However, with the ease of mathematics that comes with energy we lose something: we lose the explanation for what's physically going on.

The idea of energy started to be widely used during the Industrial Revolution to calculate how steam engines worked. It was thought that there really was a magical substance called 'Caloric' that flowed around the steam engine changing energy from one form to another.

We now know that's a fairy tale, there *is* no Caloric, and yet oddly enough the mathematics that describe how energy changes from one form to another still works as if there *was* Caloric. And engineers still use the old Caloric language, saying that energy 'flows' from one place to another.

So is the whole thing nonsense? In some ways perhaps, and yet though there is no Caloric, there often really is something else flowing, such as the hot gas flowing through a rocket or electrons flowing through a metal. The individual particles interact with one another mainly through collision, and transfer energy as they do so.

Storing energy

Another thing about energy, it can be stored for later use.

Let's look at a rechargeable battery: when we charge it up we take electrical energy and force it to flow into the battery by using electrical pressure which is called voltage and is measured in Volts. The energy flow is a real flow: a flow of electrons, which are constituent parts of atoms. Akin to mass flow rate, there is an electron flow rate which is called current, and is measured in Amps (or milliAmps which are one-thousandth of an Amp).

We've stored electrical energy inside the battery. Stored energy is called **potential energy**, because it has the potential to do something useful later on.

On the side of the rechargeable battery there will be a number followed by the letters 'mAh' as shown here which stands for milliAmp hours. This tells us how much energy can be stored inside: the bigger the number, the more electrical energy we can get out of the battery later. For example 2500 mAh means the battery can supply 2500 milliAmps (2.5 Amps) for one hour at the voltage written on the battery, or 5 Amps for half an hour.

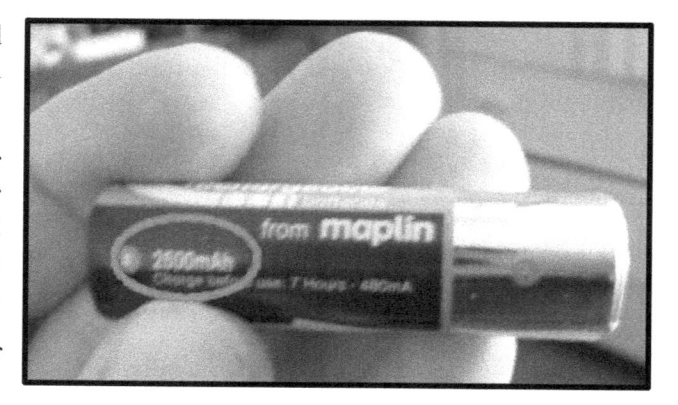

Actually, energy is measured in Joules, after the scientist James Joule. (To work out how many Joules your battery has: divide the milliAmp hours by 1000 to get Amp hours, then multiply by 60×60 =3600 to get Amp seconds, then multiply by the voltage, which on many rechargeable batteries is 1.2 Volts, or 1.2V)

Gravitational potential energy

It turns out that objects can have potential energy just because of how high up they are within the gravity field of the Earth.

For a classic example, Humpty Dumpty had potential energy because of his height upon the top of the wall. When he fell off the wall, this potential energy was turned into kinetic energy (the energy of movement, see below) because gravity speeded him up until he hit the ground.

$$\textit{specific potential energy} = \frac{\mu}{R_E + h} \quad \textit{equ. 1.20} \quad \textit{(in kiloJoules per kilogram)}$$

where μ = 398,600,500 km³s² is the Earth's gravitational constant, R_E is the mean radius of the Earth (6378.14 kilometres), and h is the height of the object above the Earth's surface *in kilometres*.

Energy of movement

If we put a battery in a toy car, we can change the electrical energy into movement energy. Movement energy is called **kinetic energy K.E.** (pronounced kin-e-tic. Recall that Kine is an ancient Greek word for moving, from which we get the word kinema or 'cinema' which is moving pictures.)

Kinetic energy depends on the speed an object has; the bigger the speed, the bigger the kinetic energy. Actually it's the velocity *squared*:

$$\textit{specific K.E.} = \frac{1}{2}V^2 \quad \textit{equ. 1.21} \quad \textit{(in Joules per kilogram)}$$

Moving objects definitely have energy because it took some energy to get them moving, and they give out this energy if they hit something.

Conservation of energy

Now here's the whole point of bothering with the concept of energy: if we put 10 Joules of electrical energy into a toy car, we'll get *exactly* 10 Joules of energy out.

So we can then work out how fast the car will end up going *without having to know how it happened*.

We don't have to find out how the electricity was stored inside the battery, nor how it got to the electric motor in the car, nor how the electric motor worked, nor how the spinning of the motor spun the wheels. Certainly some of the energy gets used in combating aerodynamic drag, and some comes out as heat energy, but we can work out these 'energy losses' quite easily.

The drag energy plus the heat energy plus the kinetic energy will exactly equal the 10 Joules put in. No more, no less, the energy is said to be **conserved** (stays the same amount) as it changes from one form to another.

So energy is an extremely useful idea because we don't have to bother to learn how things work; it saves a lot of time and physics.

Specific heat capacity and specific heat of vaporization

We know from experience that if we supply heat into an object, its temperature will rise. This relationship is described by a value known as the **specific heat capacity** 'C'. We know from page 4 that 'specific' refers to measurement per kilo, therefore Specific heat capacity is the heat required to raise one kilo of a substance by some set rise in temperature, such as one degree Kelvin. The units of specific heat capacity are therefore Joules per kilogram Kelvin.

With gasses, there are two differing experiments that can be performed to measure Specific heat capacity: we could heat/cool the gas whilst maintaining its pressure constant, or we could heat/cool the gas inside a container of fixed volume. These experiments yield two different values: Cp is the specific heat capacity of the gas if heated/cooled at constant pressure, whereas Cv is the specific heat capacity of the gas if heated/cooled at constant volume.

[Cp and Cv change slightly with temperature if the gas is not quite an Ideal gas.]

If we divide Cp by Cv we get a number known as the **ratio of specific heats** of the gas (also called the **isentropic exponent**):

$$\gamma = \frac{C_p}{C_V} \quad \text{equ. 1.22}$$

This number crops up a lot in rocketry equations. (Some rocketry books give it the symbol k.) γ gives an indication of how much energy can be stored in each gas molecule: the lower the number, the more energy can be stored.

If we subtract one from the other, we get the Specific gas constant R which we met in equation 1.14:

$$R = C_p - C_v \quad \text{equ. 1.23}$$

[As well as Specific heat capacities, we can also get Molar heat capacities, which are the values per mole rather than per kilo.]

So if we heat a substance, its temperature will rise, unless a phase change occurs. If we melt a solid, or vaporize a liquid, heat is required to alter the substance's molecular structure from solid to liquid or liquid to gas. This **phase change** occurs at a constant temperature.

The heat required to melt one kilo of substance is known as its **specific heat of fusion**, (or **latent heat of fusion**) whilst the heat required to vaporize one kilo of substance is known as its **specific heat of vaporization** (or **latent heat of vaporization**). Units are Joules per kilogram.

[When a substance is vaporized at a constant external pressure of one atmosphere this is known as 'boiling'].

Heat and internal energy

In rocketry, looking at these energies as they move though the rocket is a branch of science called **thermodynamics** from the ancient Greek words for 'heat', 'force', and 'power'.

For example, 'heat' energy stored inside the gas within the combustion chamber is called **internal energy**; an internal store that can be used later.

Now with thermodynamics, you have to be very careful with your phraseology: Although internal energy is heat, *it's not called heat* by scientists and engineers. They very carefully only use the word heat to describe heat *that flows out of the gas and into something else* such as the walls of the combustion chamber. Heat that stays internal to the gas is instead called internal energy (but yes, it is an internal store of 'thermal energy' that we can tap into).

Thermodynamic systems

In thermodynamics, the internal energy of a 'system' is energy contained within the system, excluding the kinetic energy of motion of the system as a whole and the potential energy of the system as a whole due to external force fields (gravitational, magnetic, etc.)

What is a **thermodynamic system**? It's a quantity of matter contained within a convenient enclosing boundary that we specify, such as the walls of the combustion chamber and nozzle throat. Energy inside the boundary is called internal energy, but realise that if we then decide to define a much smaller boundary, then some of that previously-defined internal energy could now be outside the boundary, so clearly isn't internal energy anymore.

So you can't say that a gas/system has a lot of heat inside it, because this doesn't make sense: heat is only energy that is flowing into or out of the system. Instead you say that a gas has a lot of internal energy. (When we say something is hot, we actually mean that it has a high temperature, *not* that it contains a lot of energy.) Internal energy behaves like what we might in everyday conversation call heat, because something that is hotter has more internal energy.

We can consider different sorts of thermodynamic systems. A useful hypothetical system in rocketry is a system that is thermally isolated from the rest of the universe, it's perfectly insulated so that no heat will flow out of it. This is called an **adiabatic** system. It may surprise you, considering how hot a rocket nozzle gets, that the flow of gas through the nozzle is *almost* adiabatic. Only a tiny fraction of the flow energy gets converted to heat through the nozzle wall. The fact that the nozzle gets so roasted shows that this is a small fraction of an enormous amount of energy!

Enthalpy and the first Law of thermodynamics

As the gas inside the combustion chamber heats up because of the high temperature of combustion, it expands (its density drops) as per the Ideal gas law (see equation 1.14). It has to do this expansion against the squashing force of the high combustion chamber pressure. So it has to work hard and use up some of its energy to do this expansion just as if it were working against a large spring. This kind of energy, struggling against a force, is called **work**. Again, we can get this energy back out later.

Work is the scientific name given to *every other way* of transferring energy into or out of a thermodynamic system *that is not heat*, so there are hundreds of different types of work: electrical work, mechanical work (moving a piston, stirring), nuclear work, etc...

The internal energy of an object (such as water) will increase if heat flows into it *or* you do work on it perhaps by stirring it. Because the internal energy increases, then so will the water's temperature.

So rather like a rechargeable battery, internal energy is a reservoir that fills or empties depending on the heat and work that fill it or drain it. This is called the **first Law of thermodynamics**, which is a version of the Law of conservation of energy. The *change* in the internal energy (ΔU, Greek letter Δ denotes finite change) of a closed system is equal to the amount of heat supplied (Q) *to* the system, minus the amount of work done (W) *by* the system on its surroundings.

$$\Delta U = Q - W \quad \text{equ. 1.24} \quad \text{(units are Joules)}$$

Now, what portion of the chemical energy went into the gas atoms shaking and moving (became internal energy), and what portion was used up as the hard *work* of expanding the gas against the strong combustion chamber pressure?

We don't need to know, because the *both* of them together soaked up all the chemical energy of combustion so we can simply add these two energies together: Engineers invented a lumped-together energy called **enthalpy** that adds these two together: enthalpy = internal energy plus pressure-work energy.

Enthalpy is given the symbol H, and as an energy, is measured in Joules (specific enthalpy h is in Joules per kilogram).

Enthalpy is generally applied to gas processes that occur at constant pressure, in this case the combustion of the propellants and their subsequent expansion occurring at constant combustion chamber pressure. (Yes, this pressure might slowly change with time, but so much more slowly than the rate of combustion.)

So if combustion released a kiloJoule of chemical energy, that turns into a kiloJoule of enthalpy, which is a potential energy that we can then use to do something useful.

(Yes, heat and light also come out of the combustion chamber gas, but these energies are miniscule compared to the enthalpy, so can be ignored.)

Both the internal energy, and the work energy done during the expansion, are connected to temperature. The hotter the gas gets during combustion (in Kelvin), the bigger both these energies get. This means that the enthalpy that results when we add them together also depends on this temperature, so enthalpy is linked to temperature:

$$h = C_p T \quad \text{equ. 1.25}$$

where C_p is the Specific heat capacity of the gas at constant pressure as discussed earlier.

The steady flow energy equation

This is just a statement of the law of conservation of energy when applied to a flow process such as the gas flowing through a jet engine or a rocket engine, both of which can be idealized as open boxes with heat (Q) coming in, and/or work (W via a turbine) coming out, through the bounding wall of the box.

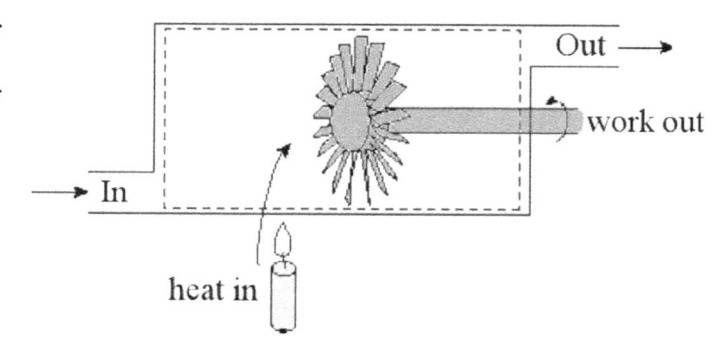

The equation deals with time rates of heat and work:

$$\dot{Q} - \dot{W} = \dot{m}\left[\Delta h + \Delta\left(\frac{1}{2}V^2\right)\right] \quad \text{equ. 1.26}$$

where the second term in the brackets $\frac{1}{2}V^2$ is the flow specific kinetic energy.

Dividing by the gas mass flow rate $\dot{m}$:

$$\Delta q - \Delta w = \Delta h + \Delta\left(\frac{1}{2}V^2\right) \quad \text{equ. 1.27}$$

where the Δ denotes a finite change between the conditions at the box inlet versus those at the box outlet.

If we take the boundary of the thermodynamic system that comprises a rocket engine to be the walls of the engine and the wall of the nozzle, then no work and (almost) no heat pass across this boundary, so $\Delta q = \Delta w = 0$. Then:

$$\Delta h + \Delta\left(\frac{1}{2}V^2\right) = 0 \quad \text{or:} \quad h + \frac{1}{2}V^2 = a\ constant \quad \text{equ. 1.28}$$

This applies within the rocket combustion chamber and all down the nozzle.

[Remember that with this selected boundary, the propellant combustion is internal to the boundary, so is internal energy, which manifests itself as a large value of the enthalpy within the box: a large value of the constant in equation 1.28 as we'll see below.]

Regarding our 'open box' boundary gives us another way of regarding enthalpy in an open flow system. Here, enthalpy is the amount of energy that is transferred across the system boundary by the moving flow. This energy has two parts: the internal energy of the fluid, and the flow work required to push the mass of fluid across the system boundary.

Stagnation conditions

As we saw earlier in equation 1.25, enthalpy is linked to temperature. But, in order to calculate enthalpy, we have to be very careful about how we envisage the temperature of the gas moving through the rocket.

As gas flows along a pipe, then from energy conservation, if the gas's kinetic energy increases for any reason, then its enthalpy has to drop by the same amount as per equation 1.28 above.

In the combustion chamber, the gas is moving very slowly (regarding it as a lump of fluid) because the chamber has a large cross-sectional area (see the mass continuity equation 1.18). So the gas's overall kinetic energy is approximately zero. If we stick a temperature probe into the combustion chamber to measure the temperature, then we'll get an accurate reading of the gas temperature in the chamber as we expect.

But if we stick a temperature probe into the fast-moving flow anywhere inside the rocket nozzle, we'll *still* get the same reading of gas temperature in the combustion chamber even though the real temperature is much lower.

What's going on?

The gas temperature is the average speed the atoms are vibrating, rotating, and zipping about, *as they're carried along by the flow*. So if you wanted to measure the nozzle flow's temperature, you'd have to use a temperature probe that was carried along at the same speed as the flow, which is almost impossible to arrange.

If you just stick an unmoving temperature probe into the nozzle flow, most of the gas will flow around this obstacle. But a tiny bit of it won't, it'll hit the probe square-on and will very briefly come to a dead stop at the upstream side of the probe. And when it momentarily stops dead it has zero speed, so all its kinetic energy vanishes and gets converted right back into enthalpy again. That's why the probe will read the combustion chamber temperature even though it's downstream of the combustion chamber.

When the little region of gas stops dead as it hits the probe, we say it is **stagnant flow**. So the temperature it measures is called the **stagnation temperature**. The flow in the combustion chamber is also nearly stagnant so that's why the probe reads the chamber temperature; it's also the stagnation temperature.

[The temperature of the *moving* flow is very confusingly called the static temperature. I don't like this phrase because the word 'static' suggests 'not moving' when the exact opposite is true. So I just call it temperature. Similarly, the pressure on the inside wall of the nozzle is called the 'static pressure' but again this is confusing; better just to call it pressure. We can in fact measure static pressure.]

Again because the flow inside it is stagnant, the combustion chamber pressure is at the stagnation pressure, and the enthalpy of the flow in the combustion chamber is the stagnation enthalpy.

Sometimes rocketeers use the word 'total' instead of the word 'stagnation', so you might see the phrase **total temperature** and so on.

Stagnation, or total, properties are generally denoted with the subscript '0'

Stagnation enthalpy is the enthalpy that would occur if the flow was stopped dead, and usefully, it remains pretty-much constant all down the nozzle as equation 1.28 says:

$$h_0 = h + specific\ K.E. = h + \tfrac{1}{2}V^2 = constant \quad \text{equ. 1.29}$$

Using stagnation enthalpy, we can rewrite equation 1.26 as:

$$\dot{Q} - \dot{W} = \dot{m}[\Delta h_0] \quad \text{equ. 1.30}$$

We can expand equation 1.27 to get the actual flow temperature T anywhere in the nozzle from a reading of stagnation temperature (see appendix 3 for derivation):

$$T = \frac{T_{stagnation}}{1+\left(\frac{\gamma-1}{2}\right)M^2} \quad \text{equ. 1.31}$$

where M is the Mach number of the flow at the point of interest, and γ is the ratio of specific heats of the gas (see page 12).

Entropy

Suppose we have a perfectly thermally insulated house (an isolated, adiabatic system) consisting of two rooms. In one room there was a heater that had previously heated the air. The other room is cold.

If we open a connecting door, then the air from the two rooms will mix, and after a time will reach a constant temperature. We say that entropy has increased with the mixing: the energy content of the house may be the same, but now it's at a 'lower quality' (quality goes down = entropy increase).

If we want to run a generator in the house powered by the heat of the air, we find that the air is now at a lower temperature in the originally 'hot' room, so our generator won't function so well. (The bigger the temperature difference between the temperature of the air and some colder region, the higher the rate of heat transferred.) It would be nice if the air then un-mixed itself, so that we could have a hot room to run our generator in, but that never happens in practice: the entropy of an isolated system never decreases.

If we apply a 'process' to a thermodynamic system, such as expanding a gas through a rocket nozzle, we find that it's an irreversible process like the mixing of the air from the two rooms, it will never reverse. However, it's worth considering a hypothetical nozzle flow that *is* **reversible**.

Reversible would mean that if you were to fix a mirror-image of the nozzle into the flow at any point down the nozzle, then you'd be able to compress and slow the flow right back down to zero speed, and you'd end up with *exactly* the same amount of enthalpy and pressure that you started with in the combustion chamber.

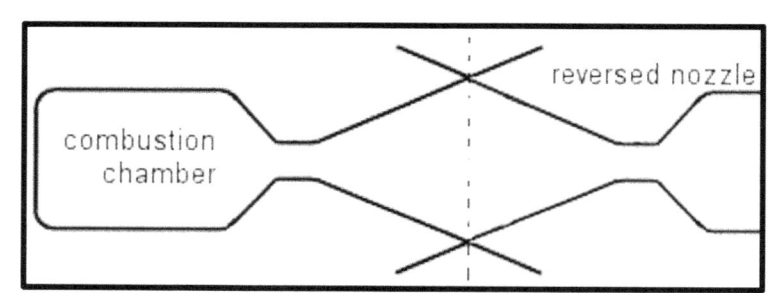

You would get all of the energy back. Getting *most* of it back but not all of it isn't good enough; the flow can only be termed reversible if you can get *all* of it back.

If the flow is assumed to be *both* adiabatic *and* reversible, then it is an **isentropic flow** (which means that entropy remains constant, it doesn't increase).

Assuming that the nozzle flow is an isentropic flow is very useful because then you can use a very helpful set of mathematical equations called the **isentropic flow relations** (see appendix 3) to let you work out the temperature drop ΔT down the nozzle just by knowing the pressure drop ΔP down the nozzle.

Is the nozzle flow really isentropic? It very nearly is, which is why the isentropic flow relations can give you a very close approximation to the correct answer for the actual properties (pressure, temperature, etc.) within the nozzle.

Pressure, temperature, density etc. are known as **thermodynamic properties**. But actually, so is entropy, because it can be derived from these three, (although there isn't an entropy-measuring device):

$$\Delta s = C_v ln\left(\frac{T_2}{T_1}\right) + R\, ln\left(\frac{v_2}{v_1}\right) \quad \text{equ. 1.32}$$

where Δs is the change in specific entropy as a gas process goes from state 1 to state 2.

v is specific volume: the reciprocal of density, R is the specific gas constant, and Cv is the specific heat capacity of the gas if it were heated in a constant-volume container (although it generally won't be). T is gas temperature. Ln is the natural logarithm function (ask your grandfather about log tables).

This equation is used to derive the isentropic flow relations, see appendix 3.

Mixture

To burn something, you collectively need three things: a **fuel** (e.g. petrol or gasoline), an **oxidiser** (some form of oxygen), and an **igniter** (something to light them with). Without all three you won't achieve combustion.

Collectively, the fuel and the oxidiser are known as **propellants**.

When you're burning propellants to get heat, it's essential to attain the correct mixture. If your mixture has too little fuel in it compared to the oxidiser (burning fuel-lean), then it won't get as hot as it could, so you'll get less performance from the rocket. Similarly, if it has too much fuel in it compared to the oxidiser (burning fuel-rich) it won't get so hot, so again you'll lose performance.

At any time during the burn, the amount of fuel being burnt divided by the amount of oxidiser being burnt is termed the **mixture ratio**

$$Mixture\ ratio = \frac{oxidiser\ mass\ being\ burnt}{fuel\ mass\ being\ burnt} \quad \text{equ. 1.33}$$

The mixture that yields the highest performance is colloquially known as **stoichiometric**, though correctly this term should be applied to the situation where the fuel and oxidiser burn at the mixture that consumes both completely: this mixture may not be the one that yields the highest performance, though it's generally close to it.

Types of rocket engine

There are three generic types of hot rockets: solid propellant, liquid propellant, and hybrid.

Solid propellant rockets

Historically, solid propellant rockets are usually referred to as rocket *motors*.

The basic solid propellant rocket is just a glorified firework, a simple combustion chamber filled with an intimate mixture of several chemicals.

Black powder is a type of gunpowder used in solid rocket motors; a powder that's been compressed until it's rock-solid and is gummed together with a binder glue. It's composed of potassium nitrate oxidiser (also called saltpetre) and a mixture of charcoal and sulphur which are the fuel.

Another popular propellant combines ammonium perchlorate 'Ap' oxidiser with one or more fuels such as powered aluminium.

Once all the propellants have been used up, the thrust stops. This is termed **burnout.**

Solids suffer from the bad combination of a **regression rate** (the speed that the burning surface erodes, producing hot gas mass) that is dependent on the gas pressure within the combustion chamber, and the fact that every exposed surface of the propellant will burn. If the flame encounters a crack in the propellant, then that's extra surface area to burn, and so the rate of gas production increases.

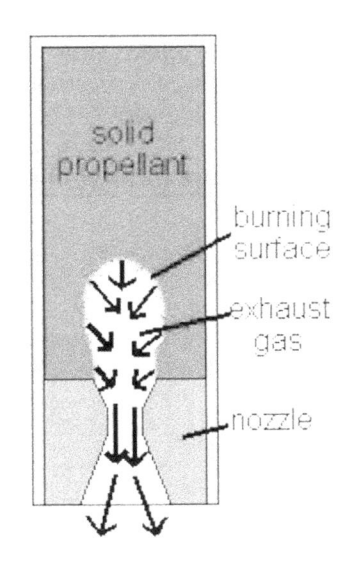

As we'll see in chapter 4, increasing gas mass inside the combustion chamber will cause the combustion chamber pressure to rise. The rising pressure causes the regression rate to increase, producing gas more rapidly, and hence the chamber pressure rises more. Meanwhile, the crack is eroding larger.

This positive feedback mechanism can be terminal: either the chamber pressure exceeds its safe limit, known as an **over-pressure** (the chamber bursts) or worse, the burning rate goes supersonic, known as **detonation**, and then the chamber *really* over-pressures!

Why is the solid fuel regression rate pressure dependent? The reason is that many solid motor fuels incorporate powdered metals, and when metallized fuels burn, the burning particles are highly emissive in the infrared, whereas the gasses produced by non-metallic fuels are fairly transparent. This infrared radiation causes an extra radiative heat transfer to the fuel grain, and its magnitude depends upon the number of metal particles per cubic metre in the combustion chamber flow (a metal density). Now, the higher the combustion chamber pressure, the higher the flow density (see the Ideal gas law) so the more metal particles that are radiating heat to the fuel grain, so the greater the radiative heat transfer.

Solid-fuel rocket motors have several advantages:

- They're simple, so there are fewer bits to go wrong.
- They don't cost as much as other types of rocket.
- There are no tanks for the propellants. (This saves weight.)
- Solid propellant is dense; it packs into the combustion chamber well, so the chamber can be smaller, and therefore lighter.

Solids do have disadvantages:

- The thrust can't be controlled during the burn.
- Once ignited, the motor can't be stopped if there's an emergency; the thrust just keeps going until burnout.
- If the propellant is damaged (has cracks in it) the motor will generally explode.
- You can't tell if a solid is about to blow up until it happens. There are no warning devices you can put inside a solid motor to warn any astronauts riding on top of it.

It's illegal to make your own solid rocket motors in the U.K. because the chemicals involved have a tendency to explode during preparation and manufacture. This is likely to maim or kill you. Fortunately, assuming you're over 18, then instead you can buy ready-made solid motors from reputable rocketry vendors.

Liquid propellant rockets

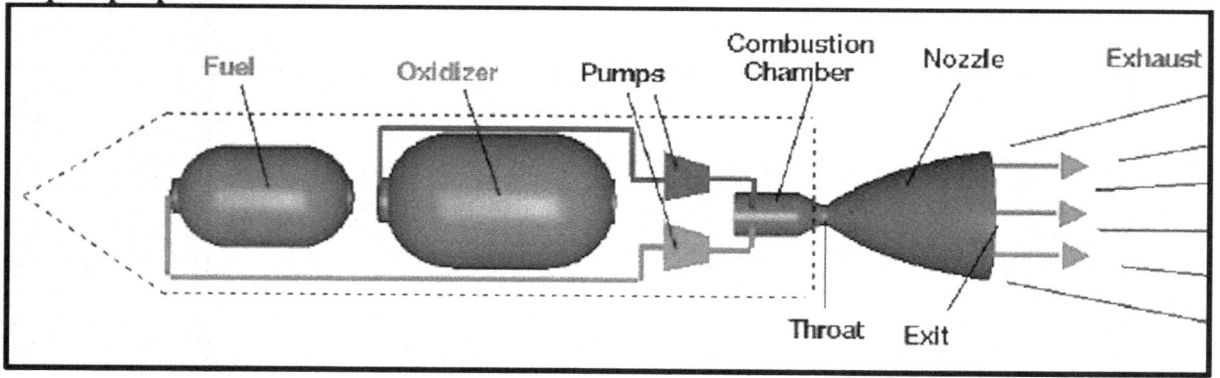

Liquid propellants remain liquid until they heat up in the combustion chamber and turn into gas. The liquid fuel and the liquid oxidiser are stored in separate tanks outside the combustion chamber.

Liquid propellant rockets that use two propellants (the fuel and the oxidiser) are also termed **bipropellant** rockets.

The fuel is often alcohol, kerosine (paraffin), or petrol (gasoline)

It's a Law of nature that fluids will only flow along a pipe if the pressure at the end of the pipe is less than the pressure at the start of the pipe. [We'll ignore liquid flowing downhill where gravity pulls it along.] The combustion chamber is pressurised to high pressure, therefore the propellants have to be pressurised to even higher pressure than the chamber so that they'll flow into it.

The propellants are pressurised by pumps, or by compressed gas fed into their tanks as per a water-rocket.

To get into the combustion chamber, the propellants flow through the **injector**. This is similar to the shower-head on a bathroom shower, as it breaks the flow into myriad streams of droplets. This is done because small droplets burn more quickly, and burn better, than one big gush of liquid does, because small droplets have a greater surface area compared to their volume.

The number of holes in the injector (known as **orifices**), and their size, determines how much propellant flows into the combustion chamber.

So by altering how many holes let the fuel in compared to how many let the oxidiser in, you can alter the mixture to get the best mixture ratio.

The injector is an expensive precision device: it has to let the two propellants through at the correct rate for stochiometric combustion, which means that its multitude of small holes must be of the correct diameter to very high tolerance to meter the flow rates accurately. Furthermore, pairs or triplets of fuel and oxidiser streams are typically aimed to splash into each other just as they enter the combustion chamber in order to create small droplets, which requires accurately drilled and polished holes, and rows of complex alternate internal channels of fuel and oxidiser.

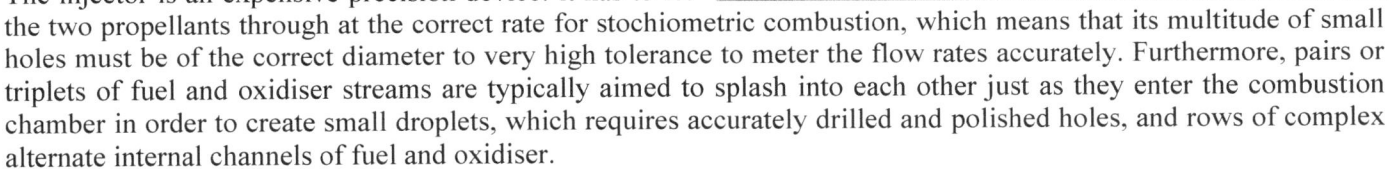

If the injector malfunctions, this is often terminal: perhaps the propellant feed system lost pressure, allowing back-flow of the hot combustion chamber gasses into the internal channels of the injector, or the injector cracked or leaked internally or melted for some other reason.

In any case, the fuel and oxidiser channels are physically close together within the injector, separated, due to restricted internal space, by thin walls. Any internal mixing, and a fire is possible which melts it all the more. Once the injector goes, the feed-pipes from the separate tanks are inevitably in close proximity, and the fire consumes the pipework back towards the tanks, which may then rupture.

Liquid-fuel rocket engines have several advantages:

- Warning devices can be placed inside the plumbing and pumps upstream of the combustion chamber to warn if something's going wrong.
- So the engine can be shut off in an emergency simply by shutting valves to turn off the flow of propellants into the combustion chamber.
- Liquids pack into the tanks better than a gas, so the tanks can be smaller, and therefore lighter than cold-gas rocket tanks.

- If pumps are used, then the liquids in the tanks can be at low pressure, so the walls of the tanks can be very thin and therefore lightweight.

They do have disadvantages:

- They're very complex (especially if they have pumps) so there are more bits to go wrong.

- Because they're so complex they cost much more to make.

- The fuel and oxidiser tanks are often right next to each other. This is dangerous if there's a fire.

- At the start of the burn you have to ignite the propellants the instant they flow into the combustion chamber. If you don't, then a puddle of mixed propellants pools in the chamber, and when you light it, it explodes, which is dryly termed 'a hard start'.

Hybrid rockets

A hybrid rocket is a cross between a solid rocket and a liquid rocket as it has a solid fuel, but a liquid oxidiser. The idea is to keep the best features of both systems whilst avoiding the worst features.

The solid fuel is inside the combustion chamber, but the liquid oxidiser comes from a tank outside the chamber. [There have been a few 'reverse hybrids' with a solid oxidiser and a liquid fuel.]

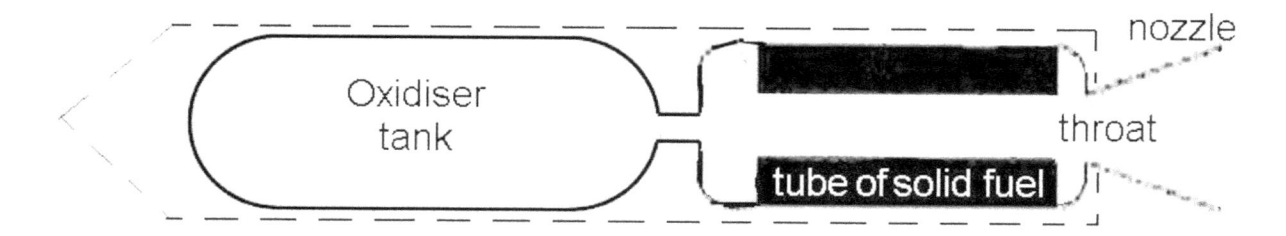

Once a hybrid is lit, the inside wall of the tube of fuel burns. This inner hole is called the **port**, which is an arcane solid-propellant rocketry term.

The rush of oxidiser down the port increases the burning, just as using bellows to blow air over burning coals on a barbecue does.

The oxidiser burns with the plastic fuel to produce a flow of hot gas down the inside of the port of fuel. But this hot gas blasting down the tube erodes a little bit of the surface of the plastic into powder which melts and burns with the oxidiser, which makes more hot gas, which makes more powder, which makes more hot gas… round and round this positive feedback progresses until settling down at a large value of hot gas flow.

The way to alter the mixture on a hybrid is to change the length of the fuel tube. The longer the tube then the longer the port, so there is more surface down the hole to farm fuel off of.

Unfortunately, hybrids change their mixture ratio during the burn. They're only at the best mix about halfway through the burn. So rocketeers choose oxidisers such as Nitrous oxide and hydrogen peroxide that give a mix that doesn't lose much performance when way off best mix.

Amateur rocketeers such as myself often make hybrids because they're safer - and much more forgiving - than liquids or solids.

This is me with the Aspirespace liquid oxygen H20 hybrid in its thrust-measuring cradle. The Lox is in the white bottle behind my arm.

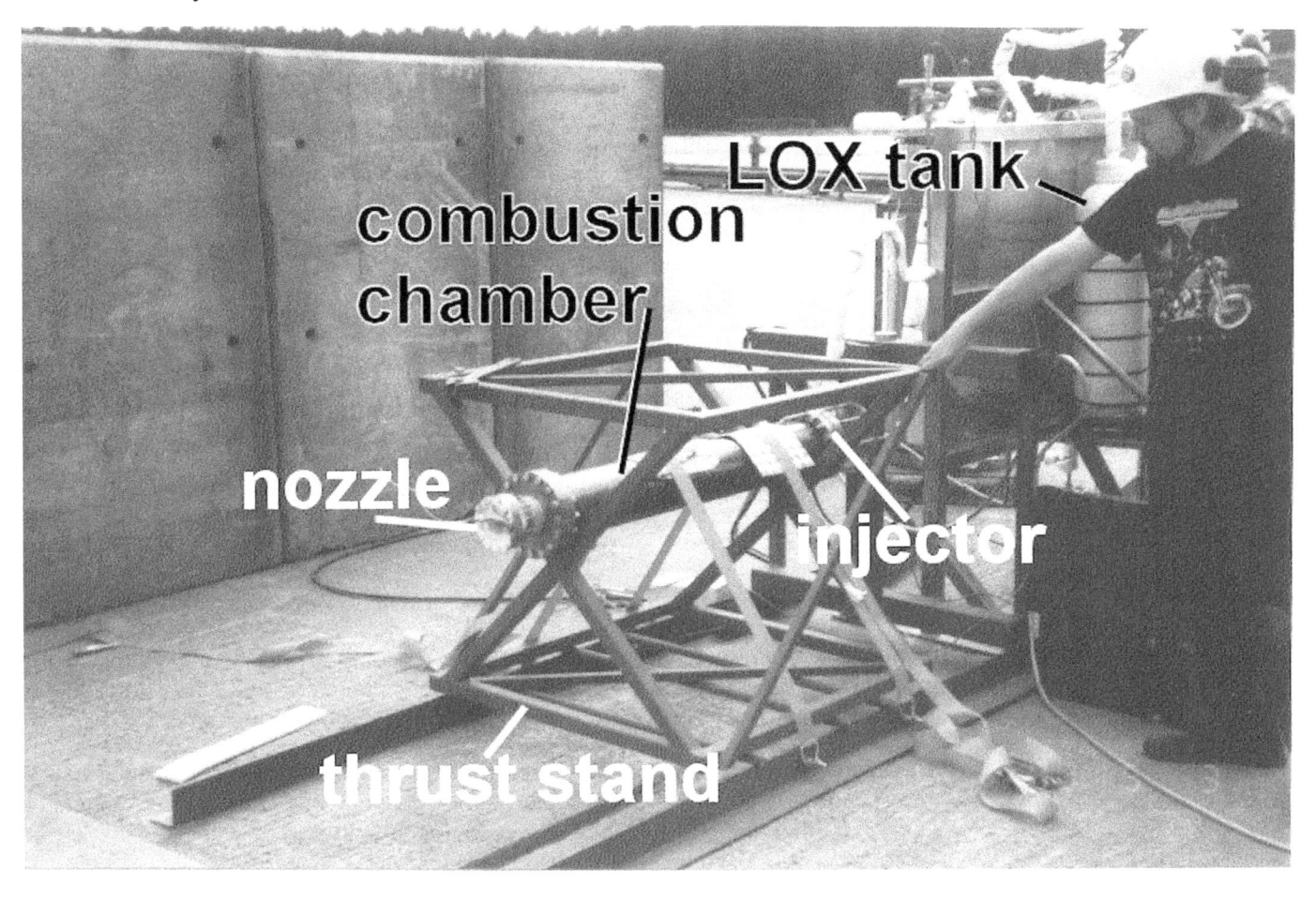

Hybrid rocket engines have several advantages:

- They can be stopped in an emergency simply by shutting a valve to turn off the flow of liquid propellant into the combustion chamber.

- Hybrids are easier to make than a liquid engine because only one set of plumbing is needed for the one liquid propellant, and only one pump is required.

- The liquid packs into the tank better than a gas, so the tank can be smaller, and therefore lighter than cold-gas rocket tanks.

- The solid plastic fuel is dense, packing into the combustion chamber well, so the chamber can be smaller, and therefore lighter.

- If a pump is used, then the liquid in the tank can be at low pressure, so the walls of the tank can be very thin and lightweight.

- The fuel and oxidiser are stored on opposite sides of the thick combustion chamber wall until they're needed. This wall stops a fire reaching both the fuel and oxidiser at the same time.

- Hybrids can't explode like liquid engines or solid motors.

They do have disadvantages:

- Hybrids are more complex than solid motors.
- It's more difficult to make very large hybrids because a large solid fuel tube doesn't burn quickly enough. This requires more than one port.

Although hybrids are safer, that doesn't make them totally safe. If they're designed badly they can still blow up.

In this book I will concentrate on hybrids using Nitrous oxide as oxidiser. Small hybrids suitable to power model rocket vehicles are known as **HPR** (somewhat optimistically called 'High power rocketry'), whereas the large rockets we will study for man-carrying vehicles are classed as **Large rockets**.

The atmosphere

We live at the bottom of a deep sea of air, which exerts a seriously high pressure on us: 101,325 Pascals, of which we have evolved to be oblivious.

The atmosphere is layered into several distinct named regions; in ascending order, the Troposphere (the turbulent region where all the weather occurs), Stratosphere (where jets fly: calm but cold), Mesosphere, Thermosphere, and Exosphere.

However, there are some overall trends that apply to the atmosphere as a whole as I'll now describe.

Step	Density
1	512
2	256
3	128
4	64
5	32
6	16
7	8
8	4
9	2
10	1

Let's climb up through the atmosphere in a series of equal steps; 10 steps for example. If we choose appropriately sized steps, and if we use suitably contrived units for atmospheric density, then to a reasonable approximation, we can generate the following table:

With each successive step (of approximately 4.9 Kilometres altitude), the density halves. The physics of the air pressing down on layers of air below generates this interesting sequence of numbers, known mathematically as an exponential decay: a step of 4.9 Kilometres is equivalent to the half-life of exponentially decaying radiation.

Interestingly, although the numbers get ever smaller, they'll never reach zero, even if you go infinitely high. So the atmosphere never really ends, it just wanders far out into Space, getting ever more diffuse.

But somewhere you've got to draw a line in the atmosphere and call everything above it Space. It's internationally recognised that you draw the line at 100 Kilometres above sea-level, as decided by the Aeronautical engineer Theodore Von Kármán: it's known, therefore, as the **Kármán line**.

A table of values of properties of Earth's atmosphere is given in Appendix 1.

A suborbital trajectory

Contrary to popular belief, if you launch a rocket on a near-vertical ascent, it will reach a certain highest point on its trajectory (known as Apoapsis or **Apogee** if above the Earth) and then simply fall back down again. The ensuing up-then-down trajectory is known as a **sub-orbital** trajectory. Only if you fire your rocket parallel to the horizon at apogee and build up sufficient sideways velocity, will you *not* fall down. This is called an orbital trajectory (generally elliptical in shape) where the 'centrifugal' acceleration of circling round the planet balances gravity.

As a rocket-vehicle ascends near-vertically from the ground, it encounters two forces: gravity of course, but also **aerodynamic drag** as it traverses the atmosphere.

The aerodynamic equation for drag is:

$$D = \tfrac{1}{2} \rho V^2 S\, C_d \quad \text{equ. 1.34}$$

where ρ is the density of the air, and 'S' is the rocket vehicle's cross-sectional area (of the fattest part of the fuselage).

'C_d', is the **Drag coefficient**. (Coefficient just means number.) The lower this number, the more easily the rocket vehicle moves through the air. This depends on the shape, which should be smooth with no sudden rough edges.

Notice that we've got $\tfrac{1}{2}V^2$ again, just as we did for the equation for specific kinetic energy. This is not a coincidence: When Daniel Bernoulli first came up with this drag equation, he worked it out using energy. He could have called $\tfrac{1}{2}\rho V^2$ 'kinetic pressure' but he didn't. Instead it's called **dynamic pressure**.

Optimum trajectory

Some suborbital trajectories are worse than others:

For example, if your rocket were to produce a thrust that is exactly equal to its weight, then it will just hover in mid-air. You won't have gained any height by the time your propellants have all been used up (burnout). You'd have lost all propellant just combating gravity without gaining any height: this is called **Gravity loss**. Of course, you'd select a higher thrust, but all ascending rocket vehicles that have a finite burn time suffer a degree of Gravity loss.

Gravity loss is expressed as a loss in velocity of the spacecraft (as an indicator of loss of kinetic energy):

$$\Delta V_{gravity\ loss} = -\int_{ignition}^{burnout} (g\sin\theta)\, dt \quad \text{equ. 1.35} \quad \text{in metres per second.}$$

g is the strength of gravity, which varies with altitude. θ is the angle of the trajectory to the horizontal and $\int(\ )dt$ is integration with respect to burn time.

To avoid Gravity loss you might think that it's a good idea to select a huge thrust, many times greater than your weight, so that the burn time is very short. But then your speed will quickly get very fast whilst you're still low down in the atmosphere. The fast speed in thick air will cause enormous drag and you'll waste a lot of propellant combating it. This is called **Drag loss**. Again it's expressed as a velocity:

$$\Delta V_{drag\ loss} = -\int_{ignition}^{burnout} \left(\frac{D}{m}\right) dt \quad \text{equ. 1.36} \quad \text{in metres per second,}$$

where D is drag and m is vehicle mass.

Using trajectory simulator software, you would discover that somewhere between these two extreme trajectories is an optimum one (highest apogee) that gives not too high a Gravity loss, and not too high a Drag loss. (And not too high gees either, you don't want to break your body.)

Indice algebra

Many rocket equations use indices (powers of) constants, usually combinations of the specific heat capacity γ.

The derivations of the equations of Appendix 3 involve juggling the indices. The algebra of indices is as follows:

$$X^a\, X^b = X^{a+b} \qquad \frac{X^a}{X^b} = X^{a-b} \qquad X^{-a} = \frac{1}{X^a} \qquad (X^a)^b = X^{ab} \quad \text{equs. 1.37}$$

Chapter 2: Propellant choices

In this chapter we'll look at the fuel and oxidiser combinations used in hybrid engines, focusing on one particular combination.

As we'll see in chapter 5, a useful measure of performance of a fuel/oxidiser combination burning properly is the specific impulse (Isp) which is crudely your 'bang for your buck': how much performance you get per kilo of propellant burnt. It's rather like the 'miles to the gallon' or 'kilometres to the litre' measure that applies to cars.

The units of specific impulse are 'seconds'.

Fuel selection

Practically anything that will burn - generally containing carbon and/or hydrogen - is suitable as a hybrid fuel grain. The full range of plastics has been tried, as well as waxes, foodstuffs, wood, and oil-soaked rolled paper.

Candy and tablet make interesting food hybrids, as well as processed foods containing fat such as salami.

Cryofuels are standard hydrocarbon liquid fuels that have been frozen, such as acetone or diesel, and even hydrogen. They have the advantage of a much higher regression rate (see chapter 7) with an insanely high rate for frozen hydrogen. Cryofuels have been extended up into the room-temperature range by using paraffin (kerosine) waxes, though very few researchers have been able to produce a mechanically robust wax grain that won't go 'gooey' when lit and spew a torrent of unburnt liquid wax out of the nozzle.

The plastics

Plastics that can be easily and cheaply moulded have a cost advantage. This has led many to experiment with rubber-like fuels such as Hydroxyl-Terminated PolyButadiene (**HTPB**) which is poured into a mould as a liquid resin. Scaled Composite's SpaceShipOne and SpaceShipTwo currently use this fuel.

However when cured, HTPB is quite soft, so it needs extra structural support within the combustion chamber. During SpaceShipOne's first spaceflight, a series of loud bangs were heard and felt at the end of the burn as broken chunks of unburnt HTPB fuel flew through the nozzle throat. Bangs so loud that the pilot thought that parts of the vehicle had broken off.

Also, rubber fuels such as HTPB create a lot of soot when burnt; give a smoky exhaust. This would cause a hybrid to dump tons of soot into the upper atmosphere. There's no rain up there to wash the soot away, so it lingers, causing an environmental problem. Frankly, the only reason HTPB ever got chosen as a hybrid fuel is because rocketeers who make solid rocket motors had lots of it lying around: it's one of the ingredients of solid rocket fuel. Granted, it performs adequately but it's hardly a very scientific choice!

There are more robust plastics, such as nylon, ABS, epoxy etc. Many of these plastics can be 3D printed, which allows the production of complex grain geometries with multiple ports.

Avoid the fluoropolymers such as Teflon because they release poisonous fluorine gas when ignited (if they ignite, which is difficult).

Here's a graph of Specific impulse versus oxidiser-to-fuel ratio for several plastic fuels burning with Nitrous oxide at sea-level. The vertical scale chosen shows an exaggerated difference between the fuels; in actual fact, their specific impulses are all within a few percent of each other as shown in the table below.

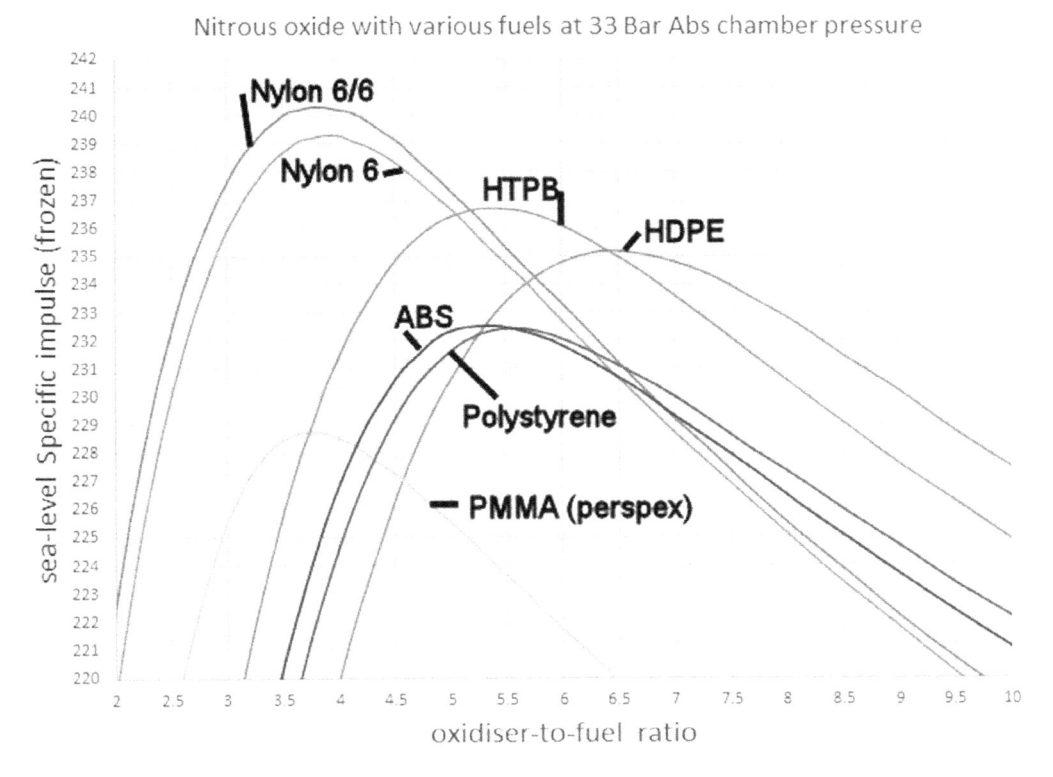

Nitrous oxide with various fuels at 33 Bar Abs chamber pressure

	Isp	Normalised Isp (percent)	Density (kg/m^3)
Nylon 6/6	240.3	100.0	1140
Nylon 6	239.3	99.6	1140
HTPB	236.7	98.5	950
HDPE	235.2	97.9	935
ABS	232.5	96.8	1070
Polystyrene	232.4	96.7	1000
PMMA	228.7	95.2	1180

From the above data, the nylons win both on specific impulse and density, though I'm not party to their regression rate characteristics.

Nylons do have one disadvantage in that their oxidiser-to-fuel ratio is quite low, meaning that proportionally more fuel has to be burnt compared to the oxidiser. As we'll see in chapter 7, this requires a proportionally larger surface area of fuel port to be provided to farm fuel off of at a sufficient rate. Balanced against this is the simple mouldability and 3D-printability of nylon which allows multiple ports to be easily fabricated.

At Aspirespace, we chose the form of polyethylene called High Density PolyEthylene, or **HDPE** two decades ago before 3D printing went mainstream. Nowadays perhaps we would switch to nylon or the similarly printable ABS. Despite the name, HDPE has a low density, requiring a larger combustion chamber to contain it.

From the above table, HDPE achieves 98% of the Isp of nylon, but its oxidiser-to-fuel ratio is almost double that of nylon, allowing us to use single circular fuel ports for reasonably-sized HPR hybrids.

This larger O/F ratio also means that the Nitrous oxidiser is the majority of the propellant mass. This is important in our spaceplane designs in that the Nitrous tank can be placed at the vehicle centre of mass (**centre of gravity**: C.G.) so that this C.G. won't wander significantly as propellant is consumed and ejected out the nozzle.

Furthermore, HDPE has double the thermal conductivity of most plastics, which should help with its regression rate.

HDPE is easy to machine but is mechanically strong: it won't break up or shatter during the burn. HDPE also burns particularly cleanly without soot, giving a transparent exhaust.

One can recycle HDPE from milk cartons then 3D print it with some effort: the inter-layer strength of the print can be weak unless care is taken with temperature control of the plastic and the build environment.

Alternatively, Robin from Celestial Mechanics has successfully fired hybrid fuel grains made from a stack of flat plastic plates laser-cut into hoops: one can buy flat sheets of recycled HDPE, so Aspirespace are experimenting with this build method. It's important to glue the grain to the chamber insulation so that there are no flow paths around and between the individual discs.

Later in this book we will select HDPE fuel, primarily because we at Aspirespace use it, so have thermochemical data for it.

The polyethylene is a long chain molecule made up of repeating chunks of C_2H_4: 2 carbon atoms and 4 hydrogen atoms. Each short straight line in this diagram represents a chemical bond, which I shall discuss fully in chapter 6.

Heating from combustion breaks the polyethylene chain into fragments that burn with the oxygen in the oxidiser.

Metal doping

In line with solid propellant fuels, many experimenters dope their hybrid fuel with powdered metals such as aluminium. This is done to try and increase performance and regression rate.

I would strongly caution against this practice for safety reasons, because you're turning the engine into a solid motor:

As we saw in the previous chapter, due to the addition of metal powders, solids suffer a regression rate that is dependent on the gas pressure within the combustion chamber. This can cause the burning to go unstable causing the chamber pressure to exceed its safe limit, known as an 'over-pressure' (the chamber bursts). In extreme cases the fuel grain can explode.

The oxidisers

The oxidisers used in rocketry are anything containing oxygen that will release this oxygen when heated sufficiently, to allow combustion.

There are several liquid oxidisers, such as Nitrous oxide, hydrogen peroxide, fuming nitric acid etc. These are discussed in our article 'Hybrid safety' on the Aspirespace website www.aspirespace.org.uk/technical_papers.html

One of the best oxidisers is **liquid oxygen (Lox)** because fuels burn fiercely in pure oxygen. But oxygen has to be chilled to **cryogenic** temperature before it will turn from a gas into a liquid. (The word cryogenic comes from the ancient Greek word kryos, meaning, 'ice cold'.) All the plumbing between the Lox tank and the combustion chamber has to be chilled to keep the Lox cold.

For reasons of safety and ease of use, we shall utilise Nitrous oxide, although Aspirespace are also developing Lox hybrids. [Safety when using Lox is all about preparation, knowledge, and absolute cleanliness of the tanks and plumbing. See the above-mentioned website article.]

Nitrous oxide

Most amateur rocketry groups choose Nitrous oxide as the oxidiser for their hybrid rocket engines. Later in this book we will design a large Nitrous hybrid rocket engine suitable of carrying one person briefly into Space.

Nitrous oxide can be referred to as 'Nitrous', and sometimes 'nitrogenous oxide' or 'dinitrogen oxide', (though not 'nitro' which is nitromethane). Nitrous is also called 'laughing gas' due to its effect on humans who are foolish enough to inhale it.

Another rocketry nick-name for Nitrous oxide is 'Nox', but 'Nox' is actually a broad environmental term for any of the various compounds and derivatives in the family of nitrogen oxides, including nitrogen dioxide, nitric acid, Nitrous oxide, nitrates, and nitric oxide.

Nitrous oxide is a molecule that has three atoms bonded together: two of them are nitrogen, and the other atom is oxygen. The Nitrous molecule is then written N_2O. When it gets hot enough, the molecule breaks up (decomposes) releasing the oxygen.

Nitrous oxide's chemical formula (N_2O) shows a predominance of Nitrogen, which doesn't help at all with combustion; it appears at first sight just to be dead-weight that has to be carried aloft, but it adds mass to the nozzle flow (which is useful as we'll see in chapter 4). It also cools the combustion chamber and nozzle (compared to pure oxygen) which stops the nozzle throat melting: you can re-use a graphite throat over several engine firings with Nitrous.

Aspects of Nitrous

1) The simple gas bottles Nitrous has to be stored in are a lot cheaper to buy or rent than, say, liquid oxygen or hydrogen peroxide containers, so at the small quantities most amateur groups use, Nitrous systems can work out cheaper, even though the Nitrous itself is quite expensive per litre.

2) Just like peroxide, a large oxidiser to fuel mixture ratio is required when burning Nitrous in the combustion chamber (around 7:1 by mass for HDPE). This results in a requirement for large quantities of Nitrous, and so a large tank onboard the rocket vehicle which is difficult to keep from being heavy.

This high mixture ratio isn't all bad, because as the oxygen within is a low fraction of the total Nitrous, you can be quite sloppy with the 7:1 *Nitrous to fuel ratio* without altering the actual *oxygen to fuel ratio* within to any degree.

This means that unlike other oxidisers, a graph of specific impulse (ISP) plotted against oxidiser-to-fuel ratio doesn't have a sharp peak at best (stoichiometric) mix that drops off sharply on either side of the peak as Lox does. The graph shown here for Nitrous/HDPE combustion (exhausting to a vacuum) is the flattest compared to the others, decreasing by less than 5% of optimum specific impulse over a range of 5:1 to 10:1 oxidiser-to-fuel ratio.

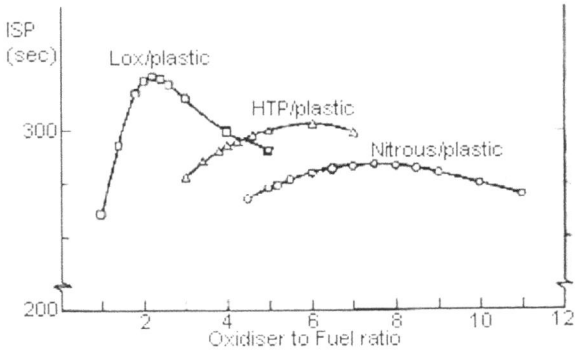

So you'll still get plenty of thrust at the beginning of your engine development programme where your mixture ratio of Nitrous to fuel might be way off stociometric.

3) Like bottled CO_2, Nitrous is subcritical at room temperature (see below) meaning that both a liquid and a vapour phase can coexist within a *closed* tank. I'll elaborate on this shortly, but the gist of it is that the moderately dense liquid phase of Nitrous can be stored in a compact tank on the launchpad in a moderate climate.

4) At room temperature, Nitrous is *only just* subcritical by a few degrees.

This is Nitrous's most unexpectedly useful property, because this close to the Critical temperature (see below), small drops in tank pressure cause large-scale production of extra vapour. This extra vapour strives to maintain the tank pressure at high value as the tank empties. A traditional blowdown system, (e.g. using an Ideal gas such as helium to pressurise the liquid), loses tank pressure at a much higher rate during the burn.

This willingness to vaporize with small pressure drops means that the Nitrous will vaporize within the orifices of even the crudest injector, typically even a simple single hole.

5) The pressure of the Nitrous gas phase (termed the **vapour pressure**) is seriously high at room temperature, at around 55 Bar (800 PSI). The gas phase can therefore be used to squirt the liquid phase into the combustion chamber at very high pressure. Some groups call this 'Vapak' pressurisation (vapour pressurisation) or 'autogenous' pressurisation.

This means you can tweak the combustion chamber to be at almost this high a pressure and the Nitrous will still run downstream (in a pressure sense) into the chamber. The higher the chamber pressure, the higher the performance of the engine, particularly at low altitudes.

6) Nitrous has to be raised to a moderately high temperature before it will decompose and release its oxygen.

This is good from a safety point of view, but it does mean that a lot of heat has to be pumped into the Nitrous from some other source at ignition, or the hybrid simply won't light-up. Once the plastic fuel is burning though, the temperature in the combustion chamber is high enough to decompose the rest of the Nitrous as it feeds in from the tank during the burn.

Subcriticality and Supercriticality

The apparent simplicity of Nitrous hybrids comes at a price because the Nitrous is typically at a temperature where its physics is not simple; it is far too cold for its vapour phase to be an Ideal gas.

Most substances, below a **Critical point** (each substance has its own Critical temperature and pressure), can exist as more than one phase simultaneously; they are then termed subcritical. Nitrous below its Critical point is termed **subcritical** whereas above it, it is termed **supercritical**.

For example: any combination of two of the solid phase, liquid phase, or gas phase, can exist together in a tank in 'phase equilibrium', or even all three at the same time at the 'Triple point.'

Nitrous oxide sitting inside a closed container at room temperature is subcritical: partly liquid, and partly gas which being *slightly* less dense than the liquid collects at the top of the container. Strictly, the term subcritical is taken to mean 'just subcritical, but near to the Critical point' but this applies to Nitrous as we'll encounter it.

Nitrous properties

Here (next page) are the variation of Nitrous oxide properties with temperature from ESDU 91022, which is worth obtaining for much more data and formulae over and above the following, which is reproduced thanks to ESDU for permission to reproduce:

Temperature (degC)	liquid density ρ (kg/m³)	vapour density ρ (kg/m³)	vapour pressure Bar Abs
-90.82	1222.8	2.7	0.9
-85	1206.7	3.8	1.2
-80	1192.7	4.9	1.6
-75	1178.3	6.3	2.1
-70	1163.7	7.9	2.7
-65	1148.8	9.9	3.5
-60	1133.6	12.2	4.3
-55	1118.0	14.9	5.3
-50	1102.0	18.0	6.5
-45	1085.6	21.6	7.9
-40	1068.8	25.6	9.4
-35	1051.4	30.3	11.2
-30	1033.4	35.6	13.2
-25	1014.8	41.6	15.5
-20	995.4	48.4	18.0
-15	975.2	56.2	20.8
-10	953.9	65.0	24.0
-5	931.4	75.0	27.4
0	907.4	86.7	31.3
5	881.6	100.2	35.5
10	853.5	116.1	40.1
15	822.2	135.4	45.1
20	786.6	159.4	50.6
25	743.9	191.1	56.6
30	688.0	237.3	63.1
35	589.4	330.5	70.3
36.42	452.0	452.0	72.5

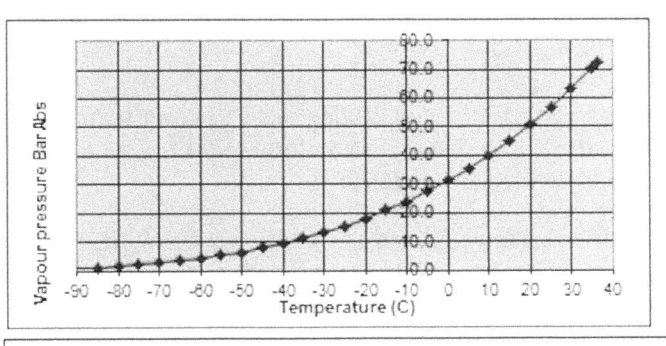

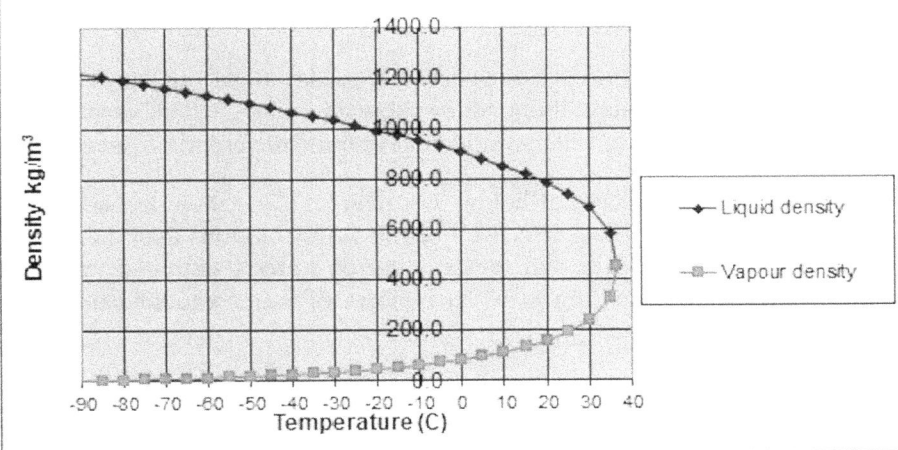

More properties used in chapter 3:

Temperature (deg. C)	Specific heat (enthalpy) of vaporization (kJ/kg)	Liquid specific heat capacity at constant pressure (kJ/kg K)
-10	255	2.079
-5	244	2.166
0	232	2.274
5	219	2.412
10	204	2.592
15	188	2.834
20	169	3.188
25	147	3.781
30	117	5.143

Here is a table of the compressibility factor (see equation 1.16) of (saturated: see below) Nitrous vapour:

Temperature (deg. C)	-30	-15	0	15	30	36.42
Z factor	0.84	0.77	0.70	0.61	0.47	0.27

Notice from the graphs above that at 36.42 degrees C, the liquid phase and vapour phase become equal in density: they merge into one phase that from then on as the temperature rises further is simply called 'a gas'.

So 36.42 degrees C is the highest temperature at which liquid Nitrous can exist. This is its Critical temperature.

At the other end of the graph, -90.82 degrees C is Nitrous' melting point: below this temperature the Nitrous is frozen solid.

You should keep your Nitrous between 15 and 20 degrees C. Lower than this temperature range, the vapour pressure is too low. Above this range, the liquid density is too low, requiring an enormous tank to contain any useful mass of it (tank volume = Nitrous mass divided by its density).

[The word 'vapour' is usually used to refer to a gas when it's below its Critical temperature and pressure, and so is existing alongside some other phase. It's purely a matter of context: there's no physical difference between a vapour and a gas, they're exactly the same thing.]

When a vapour and a liquid coexist in the one tank they are said to be **saturated**.

Living at the bottom of Earth's atmosphere as we do, all of our experience of phase changes - usually of water - occur with a constant pressure of one atmosphere around us, which usually swamps the results of our experiments. If the atmosphere wasn't there, water would behave quite differently from our usual experience.

To start with, water is subcritical below 374 degrees C so there are always at least *two* phases present below this Critical temperature. One phase may well be much less obvious than the other, in fact it's only when the temperature has climbed to 100 degrees C that the pressure of water's saturated vapour phase gets as high as the atmosphere around it. What we call boiling is when bubbles of water vapour can exist without getting squashed flat by the pressure of the Atmosphere.

So though we're used to thinking that only liquid exists below 100 degrees, and only gas above 100 degrees, this is actually a high school science simplification.

Because Nitrous goes supercritical at 36.42 degrees C, it's easy to overheat it into supercriticality. In the heat of the desert rocket launching campaigns in America, the Nitrous in several hybrids went supercritical. Supercritical Nitrous requires special injector design, so almost all thrust was lost using the standard injectors. More to the point, supercritical Nitrous violently decomposes with ease. Several supercritical Nitrous hybrids have exploded.

Here's a 3D graphical representation (not to scale) known as a **phase diagram**, of the physical properties of any substance that expands on melting, such as Nitrous oxide.

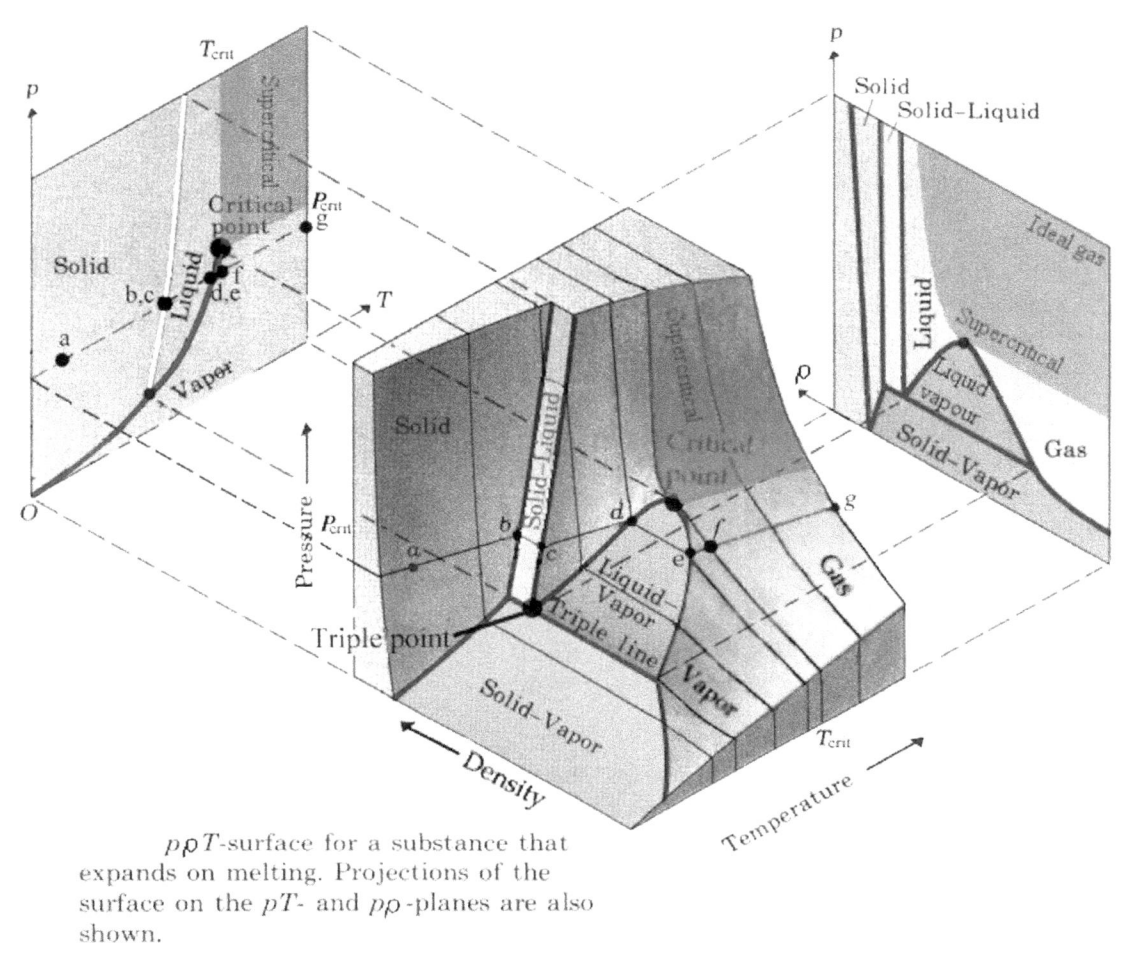

$p\rho T$-surface for a substance that expands on melting. Projections of the surface on the pT- and $p\rho$-planes are also shown.

The slopes of this chunk of 'mount Nitrous' represent the values that Nitrous physically can exist as; pressure being shown as height.

The path a-g on the upper diagram shows the 'isobaric path' (constant pressure contour), i.e. the experience we're used to with water under the constant pressure of one atmosphere around us as described above. b-c and d-e show the sudden changes of phase at constant temperature that we're used to.

[Actually, water is one of the few substances that contracts on melting, so water's phase diagram has 'c' at a higher density than 'b'; its solid-liquid 'cliff' faces away from us instead of towards us as shown for Nitrous, but in all other respects the shape of the 'water mountain' is the same.]

On our planet, Nitrous's vapour pressure is well above the pressure of the atmosphere at the temperatures we use it at: Boiling point for Nitrous is about minus 90 degrees C.

In this diagram we zoom in on the range of pressures and temperatures we'll encounter in rocketry. The density graph shows the view from 'above':

The liquid/vapour area describes what's happening in your tanks: a subcritical region where both the liquid and vapour phases coexist.

When heated, the liquid phase of Nitrous follows the saturated liquid line on the graph whereas the vapour phase follows the saturated vapour line.

The series of parallel lines (parallel to the density axis) that cross lines X and Y are known as **tie-lines**, and it's a *convention* to represent how much mass of each phase there is (as a fraction of the total mass in the tank) by the position along the tie-line.

So by this *convention* (each phase *actually* follows its respective saturation line), the exact path up the coloured section depends upon what fraction of the mass of the substance was in the form of each phase when you started heating it:

For example, path X would be a tank of Nitrous mostly filled with liquid, whereas path Y would be a tank of Nitrous with mostly vapour in it.

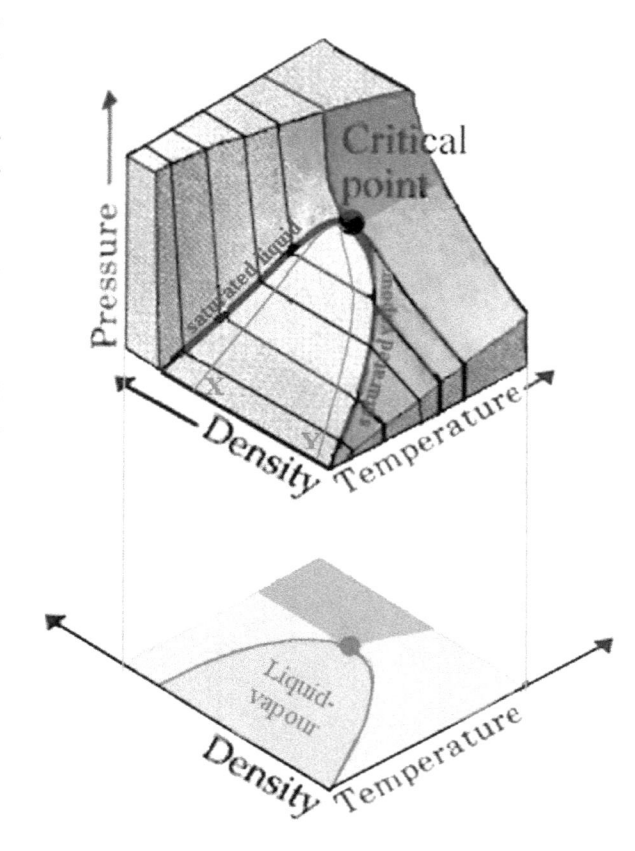

By this *convention*, the liquid saturation line is therefore the path of a tank completely full of liquid that is warming up, whereas the vapour saturation line is the path of a tank completely full of vapour.

(In the previous diagram, lines b-c and d-e are tie-lines.)

Notice that as the temperature increases, the density of the liquid saturation line decreases while the density of the vapour saturation line increases (see the table above): at the Critical point, the densities become the same; the two phases merge into one single phase, so paths X and Y both pass through the Critical point.

Changes in liquid/vapour proportion due to temperature alone

In the above diagram, look closely at the tie-lines, recalling what they represent, and you'll notice something odd about the paths X and Y.

The ratio of liquid to vapour within a closed tank changes with temperature.

This means that the amount of *liquid* Nitrous that you *think* is in your vehicle's tank will change over time if you don't take care to keep its temperature constant between the time that you *start* filling and the time that you launch. So while it may at first seem a good idea to pre-chill the tank to get a good fill of dense liquid phase in there, after several minutes the Nitrous has warmed up and so the situation has changed.

The Critical point

These are photos taken through a tank window as a substance is heated (left to right) to its Critical point. The line of the liquid surface (the meniscus) disappears: the phases merge completely.

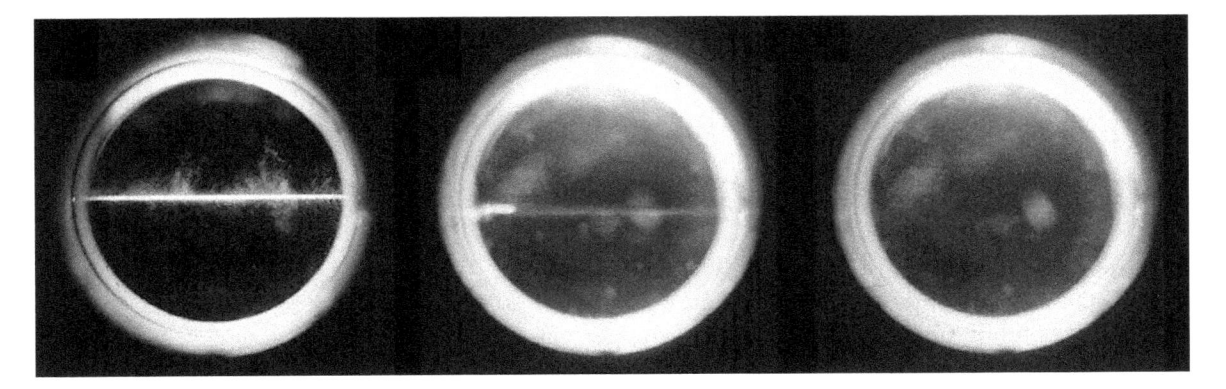

Supercritical Nitrous can therefore be regarded as either a super-dense gas, or a low density liquid.

At much higher temperatures, the density of supercritical fluid drops much lower: oxygen or nitrogen at room temperature are very supercritical, hence we refer to them as Ideal gasses (see page 8) at these conditions.

Looking at the density versus temperature diagram, you can also see that the change in density of both phases of Nitrous per degree change in temperature is largest (steepest) just before the Critical point. It turns out that the change in vapour pressure per degree C is also largest just before the Critical point.

For Nitrous, even the Scottish/Canadian climate is still rather close to its Critical temperature of 36 Degrees C, so sadly, you suffer big changes in pressure and density with *small* changes in temperature: _two Bar decrease in vapour pressure per degree C decrease in temperature_ is typical in north America/northern Europe.

This close to the Critical temperature, the Nitrous vapour phase is actually moderately dense and can't be ignored; it has a sizable mass inside the tank (and inside the combustion chamber eventually). Conversely, the liquid phase isn't terribly dense compared to other liquids, and is progressively less dense at it is warmed: heat it too much and you won't get as much mass of liquid in the tank's internal volume.

Historically, it is the liquid phase that is used in the rocket's combustion chamber. The vapour phase then causes extra thrust after the liquid runs out, but its lower density means that the burning is considerably fuel-rich, so the extra thrust it gives is less.

Nitrous slosh

Another benefit of Nitrous oxide is one I've never heard mentioned anywhere else and that is to do with a phenomenon called **slosh mass**. In traditional liquid propellant tanks, when the vehicle is oscillating in pitch, a wave of propellant sloshes from side to side on the top surface of the propellant in its tank.

For small model rockets this isn't a problem, but as the vehicle gets appreciably larger, the slosh wave can get physically heavy; so much so that it can upset the vehicle's orientation control system if the control system frequency happens to coincide with the same frequency of this wave sloshing from side to side.

With Nitrous oxide, as the tank is emptying, the mass of the Nitrous vapour on top of the Nitrous liquid is actually quite heavy, so when a wave of liquid Nitrous sloshes one way, an anti-wave of Nitrous vapour sloshes in the opposite direction and reduces the overall effect of the sloshing. As the frequency of these sloshing waves depends on their fluid density, and the density varies over the burn, that's a whole range of frequencies that the control system has to avoid.

Chapter 3: Using Nitrous oxide in rocketry

Nitrous oxide, like any just-subcritical propellant, has to be handled in particular ways during hybrid filling and operation. The first step is decanting it from the sturdy bottle you bought it in, known as the **fill-tank**, into the rocket's own propellant tank, known as the **run-tank** because it's connected to the hybrid when the engine is running (firing).

Filling the run-tank

The run-tank is filled using a difference in pressure between the fill and run-tanks:

The pressure difference (or **head**) can be caused by gravity when the run-tank is connected to a fill-tank that is physically higher than it. This will fill the run-tank with the denser liquid phase, while the slightly lighter gas phase will bubble back up into the fill-tank.

Alternatively, the run-tank has a vent-hole in it which is open to the atmosphere. This lowers the pressure in the run-tank relative to the fill-tank (see diagram on next page).

Then the large pressure difference between the inside of the fill-tank and the outside air will carry the Nitrous several metres 'uphill' into the run-tank. So the fill-tank can then be physically lower than the run-tank; it is typically lying on the ground whilst the run-tank is up in the rocket vehicle up on the launcher.

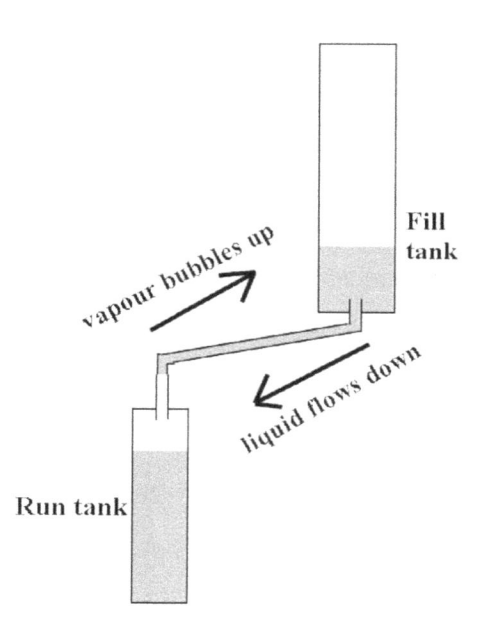

The fill

The filling process works thusly: at the beginning of the fill, the run-tank is at ambient pressure, and is full of air. Then as the fill proceeds, liquid Nitrous pours into the run-tank, but the low tank pressure causes the liquid to vaporize completely into vapour.

Due to a small vent hole at the top, Nitrous vapour gradually fills the run-tank, expelling the air out the vent, gradually increasing the pressure in the run-tank until the pressure causes the vapour to condense into liquid. The liquid falls to the bottom of the run-tank and collects, with the less dense vapour above. Tests suggest that it only takes a few seconds for the run-tank to reach *almost* pressure equilibrium with the fill-tank, after which the fill begins in earnest.

How full?

A question often asked is, "how full can you fill the run-tank?"

As we'll see later, the expulsion of the liquid Nitrous phase out of the tank during the burn *is not* a simple blowdown process, because the Nitrous vapour is definitely not an Ideal gas.

[A blowdown system uses a fixed reservoir of Ideal gas such as nitrogen to force the liquid propellant out of the run-tank.]

Nitrous performs much better than this, and in fact test-firings and simulations show that the graph of tank pressure drops with time (during the firing). It *does not* depend upon the amount of Nitrous vapour originally in the top of the tank, so you could fill the tank completely full of liquid.

But if the run-tank is to be completely sealed after filling, but then left for some time before firing, then for safety reasons, a small percentage of the tank volume should be deliberately left free of liquid to allow for liquid Nitrous expansion with the ensuing increase of temperature. This **ullage** (see chapter 8) is a space at the top of the tank that is filled with Nitrous vapour instead of Nitrous liquid. It's also called a head-space.

Vents

On many hybrid systems, ullage is achieved by a vent-hole or vent-pipe with an inlet situated slightly below the top of the tank; the outlet is open to the atmosphere outside.

A vent works exactly like the overflow outlet on a bathtub in that the liquid never fills higher than the vent. (Provided that you fill it reasonably slowly.)

The outlet of the vent-pipe can be higher than the vent inlet if required, because the large pressure difference between inside the tank and outside will happily carry the Nitrous several metres 'uphill'.

As soon as the liquid Nitrous reaches the level of the vent, you'll see the plume issuing from the vent thicken and whiten appreciably, and that's the time to stop filling. A dark background behind the vent outlet aids this visual check.

If your hybrid design allows, now's also the time to close the vent hole to stop the loss of Nitrous. With our large hybrid, this will be essential.

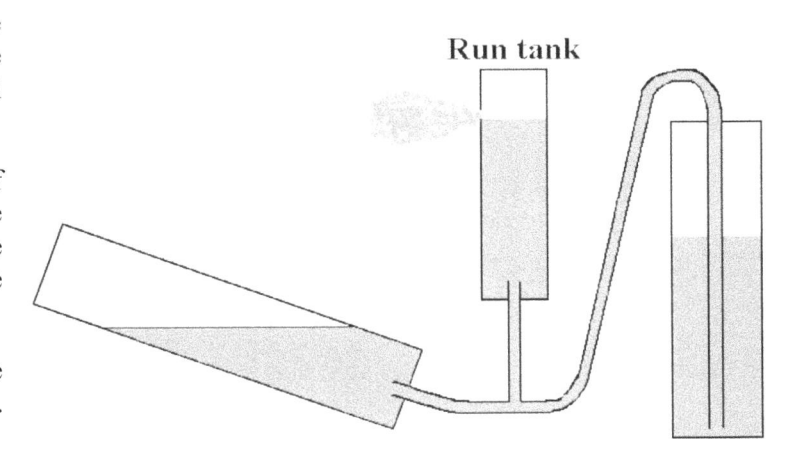

Most commercial HPR Nitrous hybrid systems keep the vent open permanently, therefore Nitrous is continuously being lost. Although a small enough vent diameter will keep the tank pressure high for some time, this progressively lowers the tank vapour pressure over time as discussed below, so such a design has to be launched *immediately* after filling. Wait too long and significant thrust will be lost.

In the above diagram, the fill-tank on the left has to be tilted-up to get liquid phase out, whereas the fill-tank on the right has a **dip-tube** (or 'siphon tube') running down inside it so that it can be sat upright. When you purchase your Nitrous, remember to ask whether the fill-tank has a dip tube fitted or not.

Changes in liquid/vapour proportion due to temperature alone

As we saw in chapter 2, the ratio of liquid to vapour within a closed run-tank changes with temperature.

This means that the amount of *liquid* Nitrous that you *think* is in your run-tank will change over time if you don't take care to keep its temperature constant between the time that you *start* filling and the time that you launch.

The Nitrous is contained inside the fixed volume of the closed bottle that is the run-tank, and so its mass can't change with time:

$$m_{total} = m_{liquid} + m_{vapour} = a\ constant \qquad \text{equ. 3.1}$$

So it's forced to self-adjust so that it can physically fit inside the tank as the densities of the two phases change with temperature.

The way it physically alters the volumes of the liquid and vapour phases is that a rise in temperature causes some of the liquid to vaporize into vapour, whilst a drop in temperature causes some of the vapour to condense into liquid.

It's forced to follow a volume formula:

$$V_{vapour} + V_{liquid} = V_{bottle} \quad \text{or:} \quad \frac{m_{liquid}}{\rho_{liquid}} + \frac{m_{vapour}}{\rho_{vapour}} = V_{bottle} \quad \text{equ. 3.2}$$

where ρ is density, m is mass.

This 'self-adjustment' phenomenon is effectively a reversible chemical reaction. Temperature is defined as the *average* speed of the molecules of the Nitrous: some are moving slower than the average, while some are moving faster, possibly fast enough to break away from the liquid phase and become part of the vapour. This is known as **evaporation**. Conversely, some of the slower vapour molecules that 'impact' the liquid phase remain as part of the liquid, a process known as **condensation**.

At equilibrium (where the Nitrous has reached constant temperature and pressure), the rate of condensation is exactly balanced by the rate of evaporation, so no net change occurs with time.

It's only when the Nitrous is no longer in equilibrium that one of the rates exceeds the other, and an overall change occurs.

This all occurs within your closed run-tank and so you can't see it happening. Worse still, the total mass of Nitrous in the bottle remains the same of course, so weighing scales won't pick up any changes in the proportion of liquid to vapour.

The following resolves this problem:

Fill calculation (after closing the vent valve)

Assuming that you filled the run-tank slowly then you know what mass of Nitrous went into the tank, because the volume of tank above the vent-hole should have been vapour alone, and the volume of tank below the vent-hole should have been liquid alone.

So for example if the ullage space was 12% of the tank volume, then just at the end of filling:

$$\frac{m_{vapour}}{\rho_{vapour}} = 0.12(V_{tank}) \quad \text{equ. 3.3}$$

and:

$$\frac{m_{liquid}}{\rho_{liquid}} = (1 - 0.12)(V_{tank}) \quad \text{equ. 3.4}$$

If you don't have weighing scales, these two combine to give:

$$m_{total} = m_{liquid} + m_{vapour} = 0.12(V_{tank}) + (1 - 0.12)(V_{tank}) \quad \text{equ. 3.5}$$

The densities of the saturated liquid and saturated vapour phases can be read off of a lookup table such as given in chapter 2.

A run-tank pressure-gauge is invaluable here, to discern what temperature caused this run-tank vapour pressure reading; it may not have reached ambient temperature yet. [As the Nitrous is 'saturated' then the Nitrous temperature and the vapour pressure are linked: the pressure varies with temperature.]

The changes in the proportions of the two phases after some time when the temperature has changed (noted by a change in the vapour pressure reading) can then be calculated by rearranging equation (3.2):

$$m_{liquid} = \rho_{liquid}\left(V_{tank} - \frac{m_{vapour}}{\rho_{vapour}}\right) \quad \text{equ. 3.6}$$

substituting equation (3.1):

$$m_{liquid} = \rho_{liquid}\left(V_{tank} - \frac{(m_{total} - m_{liquid})}{\rho_{vapour}}\right) \quad \text{equ. 3.7}$$

collecting terms:

$$m_{liquid}\left(1 - \frac{\rho_{liquid}}{\rho_{vapour}}\right) = \rho_{liquid}\left(V_{tank} - \frac{m_{total}}{\rho_{vapour}}\right) \quad \text{equ. 3.8}$$

dividing by ρ_{liquid}:

$$m_{liquid}\left(\frac{1}{\rho_{liquid}} - \frac{1}{\rho_{vapour}}\right) = \left(V_{tank} - \frac{m_{total}}{\rho_{vapour}}\right) \quad \text{equ. 3.9}$$

giving:

$$m_{liquid} = \frac{\left(V_{tank} - \frac{m_{total}}{\rho_{vapour}}\right)}{\left(\frac{1}{\rho_{liquid}} - \frac{1}{\rho_{vapour}}\right)} \quad \text{equ. 3.10}$$

and:

$$m_{vapour} = m_{total} - m_{liquid} \quad \text{equ. 3.11}$$

where the densities are those at the new temperature, and m_{total} and V_{tank} have of course remained constant.

Proportion changes due to outflow

Because pressure, temperature, and density are connected, if we cause changes in *pressure* within our run-tank, either during filling, or when we empty its contents into the combustion chamber, temperature changes will then occur. And as we've seen, temperature changes cause the ratio of liquid mass to vapour mass in the run-tank to change.

Several examples of this occur during hybrid operation:

Firstly, the vent-hole relies on the fact that the vapour pressure inside the run-tank is higher than the atmosphere outside, and so an outflow is established.

The vent should either be of tiny diameter, or be a pipe with a restriction of tiny diameter somewhere along it. (0.3 millimetres diameter is typical.)

A large diameter vent is undesirable because it provides little resistance to the flow pouring out of it, so the drop in pressure between tank and outside occurs more within the tank than within the vent hole. [Electronics engineers will recall the principle of a Potential Divider.]

The Nitrous responds to this low tank pressure by vaporizing its liquid away large-scale. Moreover, the flow rate of Nitrous leaving via the vent-pipe is much higher, so it'll all disappear after a short time. Also, a vent produces gas thrust like any rocket, so you want this flow rate to be small if it's venting sideways.

Similarly, when the **run-valve** opens, (the valve between run-tank and combustion chamber) the gas phase forces the liquid phase out of the tank in the manner of a water rocket, because the combustion chamber connected below the tank is also at lower pressure.

As the tank empties, the liquid level obviously drops, so the volume available to the vapour phase above the liquid increases, so the vapour expands. And like any gas, its pressure drops as it expands.

Whatever caused the vapour phase's pressure to drop - be it venting or emptying - the pressure is now lower than it ought to be (it ought to be at the vapour pressure corresponding to its temperature). This drop in pressure is 'sensed' by the liquid phase below it.

Some of the liquid phase will then vaporize in an attempt to create more vapour to raise the tank pressure back up to vapour pressure: the lower the pressure (the bigger the pressure imbalance), the higher the vaporization rate.

Now the process of vaporizing liquid into vapour requires energy (called the latent heat of vaporization, see page 12), and this energy has to come from somewhere. The energy required is drained from the nearest available source, which in this case is the remaining liquid Nitrous itself, which therefore gets cooled by an amount determined by its specific heat capacity (see page 12).

Oddly enough, my experiments and simulations show that the metal wall of a Nitrous tank doesn't give up heat that quickly into the liquid even though you'd expect it to: the metal may be a conductor, but the liquid interface between the metal and the bulk of the liquid isn't. So the metal of the tank can be ignored as a heat source for pressure changes, *provided* that they occur in a short time, say the 10 seconds or less that are typical of a small HPR hybrid firing.

There will be thermal layering effects (known as **stratification**) occuring within the Nitrous: wherein the liquid and vapour closest to the boundary between liquid and vapour ought to be the coldest because that's where the vaporization occurs.

However, from an engineering point of view, the stratification can be completely ignored in computer modelling.

Perhaps this is because the colder Nitrous will be denser, so will try to sink to the bottom of the tank and so the liquid gets evenly mixed. Also, changes in pressure affect the whole of the Nitrous at once.

Experiments show that other effects cancel stratification out, and so the liquid and vapour can be simply modelled as 'blocks' at uniform temperature. (See section 'Modelling the Nitrous run-tank emptying' at the end of this chapter.)

This cooling of the remaining liquid (and therefore any future gas to be vaporized from its surface as the emptying progresses) means that the vapour pressure (the tank pressure) will slowly drop over the burn time. In this graph, burnout was taken as the point when the liquid phase ran out (the graph suddenly steepens):

The lower the pressure drops below vapour pressure, the more vapour is required to raise the pressure back up, and the more chilled the liquid phase becomes as it provides this vapour.

This is why leaks in any pipe-joints carrying the liquid phase of Nitrous oxide show up as regions covered in ice; the Nitrous sucks heat out of the atmosphere as it leaks out to atmospheric pressure and vaporizes. This freezes the water vapour in the air around the leak.

So if you crank open the vent (to the atmosphere outside) to huge diameter in an attempt to perform a quick fill, you'll lower the tank pressure *way* below vapour pressure. So the Nitrous will vaporize large-scale, chilling itself seriously cold in the process as it drains heat from itself.

If the leak is plugged, for example by shutting a valve on the vent-line, or by shutting the **run-valve** mid-burn, liquid will continue to vaporize inside until the vapour pressure is restored. (Albeit the lower vapour pressure you get at a colder Nitrous temperature.)

Then as heat from outside *slowly* trickles back into the liquid through the tank walls (this takes a long time, so the tank *does* count as a heat source), the vapour pressure will slowly rise again until the liquid is back at ambient temperature, then no more heat can flow in. This can take a good 15 minutes even for small run-tanks though.

The rate of decrease of tank pressure with time (the slope of the graph above) depends on how quickly you empty the tank:

Experiments at Surrey Satellites Technology Ltd have shown that if the Nitrous is emptied at a tiny flow rate, less than 10 grams per second or so, then the tank pressure remains constant because the small inflow of heat through the tank wall is just enough to compensate.

Boiling

The vaporization of the liquid phase into gas is known to resemble conventional 'boiling':

Analogous to the phenomena of supercooling, the boiling of water at atmospheric pressure sometimes doesn't occur at the boiling temperature of 100 degrees C. Sometimes the temperature continues to rise higher until some tiny dent or scratch in the container wall (called a 'vapour nucleation site') forms a bubble that breaks loose and sets the wholescale boiling off.

Chemists often drop 'boiling stones' (small porous bubble-producing 'pebbles') into beakers to ensure that boiling occurs at the temperature expected.

Experiments show that mechanical agitation will also trigger boiling in fluid that ought to be boiling but as yet is not.

Once any tiny amount of local boiling kicks in, the resulting bubbles agitate the liquid, greatly increasing the boiling rate, and this feedback mechanism then cascades to produce serious boiling.

We see this in our hybrid tests too: the graph shown on the previous page is typical, and shows a downward kink at the start.

It seems that when we open our hybrid **run-valve**, the initial drop in the liquid level catches the Nitrous 'unawares', and so there is very little vaporization: the ensuing graph of pressure drop has the characteristic steepness of an expanding Ideal gas.

Once the hybrid fires up though, the ensuing engine vibrations shake the rocket vehicle, and hence the tank. This shaking appears to trigger large-scale 'boiling', and the tank pressure graph then rises sharply, before descending at the gentler slope that you'd expect from a vaporizing subcritical fluid.

The vapour-only phase

After all the liquid Nitrous has run out of the run-tank, there will still be some vapour remaining. Even if you started with a tank completely full of liquid, some vapour will have been created as the tank emptied.

This vapour is dense enough to erode the hybrid fuel grain and so produce thrust, though it burns fuel-rich (too little oxidiser) which lowers the performance (the Specific impulse, chapter 5). From our hybrid firing data, we've learned a few surprising things about this 'vapour-only' phase of combustion:

- It transpires that the pressure loss that occurs as the vapour flows through the injector orifices is identical to when the liquid was flowing through it. This proves that the liquid vaporizes completely to vapour inside the orifices because according to Bernoulli's principle (see appendix 3) the flow velocity causes a pressure drop (flow velocity goes up, pressure goes down).

- The vapour emptying out of the run-tank very nearly follows an isentropic process (see page 16). That means that very little energy is wasted (negligible increase of entropy) during the emptying, and no heat is transferred from the tank walls to the vapour.

- Therefore the vapour pressure and temperature drop rapidly as the tank empties and the vapour expands.
- The vapour is not an 'Ideal gas'. Intermolecular forces (the forces between the vapour molecules) are noticeably at work, so Nitrous vapour expands differently to that of an Ideal gas. Use a compressibility factor (chapter 1).

With the above in mind, a simple mathematical model will simulate the tank emptying as detailed below, the results of which are shown here:

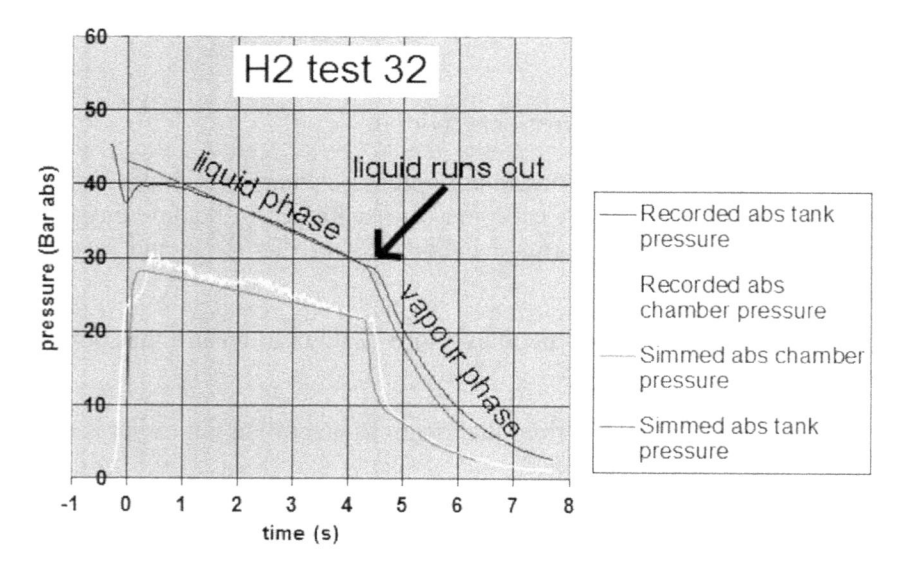

[Graph: H2 test 32 — pressure (Bar abs) vs time (s), showing liquid phase, "liquid runs out", and vapour phase, with legend: Recorded abs tank pressure, Recorded abs chamber pressure, Simmed abs chamber pressure, Simmed abs tank pressure]

We'll investigate the vapour phase burn more fully in chapter 9, where we'll discover that its performance is very dependent upon the pressure of the atmosphere outside the engine nozzle: at very high altitudes where this pressure is low, its performance is adequate.

Vapour burn time

In a Nitrous hybrid, the ratio of liquid phase burn time to vapour phase burn time is roughly 3:2, though again, the vapour phase burn time does depend on the ambient (atmospheric) pressure outside the nozzle.

Modelling the Nitrous run-tank emptying

The following mathematical model of the liquid phase emptying is based on a model of saturated propane emptying from a run-tank devised by Dr. Bruce Dunn (see references).

When the Nitrous vapour expands due to the level of the liquid Nitrous dropping, the vapour pressure drops due to this expansion.

We don't need to know what this pressure drop is to model the tank emptying, instead we estimate how much mass of Nitrous liquid, m_V, has been vaporized to try and raise the pressure back up again to its original value.

We'll break the tank emptying process into a sequence of short time intervals, $\Delta time$, and calculate the Nitrous mass remaining in the tank after each time interval. This is an iterative process; we pick an arbitrary non-zero value for m_V to start with, and the program quickly converges on the actual value, and stays with it as it changes as the tank empties.

We calculate the heat removed (ΔQ) from the liquid Nitrous during its vaporization:

$\Delta Q = m_V H_V$ equ. 3. 12

where H_V is the enthalpy (latent heat) of vaporization evaluated at the current Nitrous temperature (how much energy is required to vaporize Nitrous per kilogram). See chapter 2 for the required Nitrous oxide data.

We then calculate the temperature drop of the remaining liquid Nitrous (m_{liquid}) due to losing this heat:

$$\Delta T = \frac{-\Delta Q}{m_{liquid}\, C_{liquid}} \qquad \text{equ. 3. 13}$$

where C_{liquid} is the specific heat capacity of liquid Nitrous at the current temperature (chapter 2).

We then subtract this temperature drop from the current liquid Nitrous temperature to get a new lower liquid Nitrous temperature.

The liquid density ρ_{liquid}, the vapour density ρ_{vapour}, and the vapour pressure (tank pressure) are now recalculated from tables of Nitrous properties (see chapter 2) based on this lower temperature.

Using this new tank pressure and the current combustion chamber pressure, the mass flow rate of liquid Nitrous out of the tank, $\dot{m}_{liquid}$, is now calculated using equation 3.13 from appendix 3 (see appendix 3 for details):

$$\dot{m}_{liquid} = N A_{orifice} \sqrt{\frac{2 \rho_{liquid}\, \Delta P}{C}}$$

Having calculated the mass flow rate of liquid out of the tank we can now integrate it to get the mass that has left the tank in this time iteration $\Delta time$:

Total system mass m_{total} (liquid + vapour) has decreased by $\dot{m}_{liquid}\, \Delta time$

Liquid mass m_{liquid} has decreased by $\dot{m}_{liquid}\, \Delta time$

The resulting value for m_{liquid} is the mass of liquid that would be in the tank if the Nitrous did not react to the expansion of the Nitrous vapour and the ensuing drop in pressure. We'll designate this as m_{liquid_old}

But the Nitrous *does* react, both to the increase in Nitrous volume and also the drop in temperature.

The densities of the liquid and vapour are functions of temperature only. The Nitrous is constrained to fit into the volume of the tank, so is forced to adhere to a volume formula (equation 3.10):

$$m_{liquid} = \frac{\left(V_{tank} - \dfrac{m_{total}}{\rho_{vapour}}\right)}{\left(\dfrac{1}{\rho_{liquid}} - \dfrac{1}{\rho_{vapour}}\right)}$$

we'll designate this value as m_{liquid_new}

The discrepancy between this value and the previous value is the mass of Nitrous that has been vaporized

$$m_v = m_{liquid\_new} - m_{liquid\_new} \qquad \text{equ. 3. 14}$$

With this new value for m_v we can proceed to the next time iteration, and begin the calculation loop again.

N.B.

Bear in mind that this model is an approximation only, which uses simple integration routines. Occasionally, the model goes awry just as the last of the liquid Nitrous is emptying. Add the following logic check to catch this, and use it to trigger engine burnout.

    if (m_{liquid_new} is greater than m_{liquid_old}) then trigger engine burnout

Modelling the vapour-only phase

Capture and store the initial tank values that occurred the instant the last of the liquid ran out of the tank exit: initial vapour temperature Ti, initial vapour mass mi, initial vapour pressure Pi and temperature Ti, and initial vapour density ρi

Then work out the initial compressibility factor Zi using Ti (see the compressibility factor table in chapter 2).

Next, calculate the mass flow rate of Nitrous vapour out of the tank using equation 3.13 from appendix 3 again as above, but working with vapour instead of liquid. The loss coefficient C remains the same.

Next, the vapour emptying can be modelled as an isentropic process (see page 16). The inter-relationships between P, T, and ρ for an isentropic process are modelled using the standard isentropic equations as listed in appendix 3:

$$\frac{T_2}{T_1} = \left(\frac{P_2}{P_1}\right)^{\frac{\gamma-1}{\gamma}} = \left(\frac{\rho_2}{\rho_1}\right)^{\gamma-1}$$

where γ is the ratio of specific heats which is 1.3 for Nitrous vapour. (Averaged over the subcritical temperature range of interest.)

We'll take time 1 as the initial value as the liquid just runs out, and time 2 as a point some time later on in the emptying process.

Starting with the gas equation for a real gas: $P = Z\rho RT$ (equation 1.16) where R is the specific gas constant for Nitrous.

Rearranging this equation, and substituting $\rho = \dfrac{m}{V_{tank}}$ where V_{tank} is run-tank volume, gives:

$$\frac{T_2}{T_1} = \frac{\left(\frac{P_2}{Z_2 m_2}\right)\left(\frac{V_{tank}}{R}\right)}{\left(\frac{P_1}{Z_1 m_1}\right)\left(\frac{V_{tank}}{R}\right)} \quad \text{equ. 3. 15}$$

Rearranging and cancelling gives:

$$\frac{T_2}{T_1} = \frac{P_2}{P_1}\left(\frac{Z_1 m_1}{Z_2 m_2}\right) \quad \text{equ. 3. 16}$$

Using the isentropic flow equations above to substitute temperature for pressure:

$$\frac{T_2}{T_1} = \left(\frac{T_2}{T_1}\right)^{\frac{\gamma}{\gamma-1}}\left(\frac{Z_1 m_1}{Z_2 m_2}\right) \quad \text{equ. 3. 17}$$

Taking the temperatures over to the left-hand side:

$$\left(\frac{T_2}{T_1}\right)^{\frac{(\gamma-1)-\gamma}{\gamma-1}} = \left(\frac{Z_1 m_1}{Z_2 m_2}\right) \quad \text{or:} \quad \left(\frac{T_2}{T_1}\right)^{\frac{-1}{\gamma-1}} = \left(\frac{Z_1 m_1}{Z_2 m_2}\right) \quad \text{equ. 3. 18}$$

Giving:

$$\left(\frac{T_2}{T_1}\right) = \left(\frac{Z_1 m_1}{Z_2 m_2}\right)^{\frac{\gamma-1}{-1}} = \left(\frac{Z_2 m_2}{Z_1 m_1}\right)^{\gamma-1} \quad \text{equ. 3. 19}$$

If we use the initial values we stored earlier as time 1, then this equation gives the temperature at time 2, when the new mass of vapour within the tank is m_2.

Now we can use the above isentropic flow equations to calculate the new vapour pressure and density at time 2.

There is one problem though: we need to calculate Z_2 in equation 3.19, but this depends upon T_2, the vapour temperature at time 2, which we calculate *after* calculating equation 3.19

So we need to resort to a recursive calculation loop:

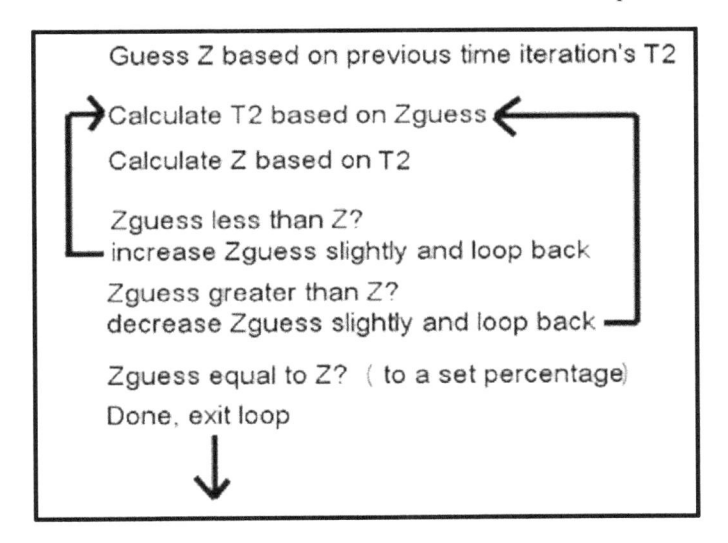

```
Guess Z based on previous time iteration's T2

→ Calculate T2 based on Zguess ←
  Calculate Z based on T2

  Zguess less than Z?
  increase Zguess slightly and loop back

  Zguess greater than Z?
  decrease Zguess slightly and loop back

  Zguess equal to Z?  ( to a set percentage)
  Done, exit loop
       ↓
```

Now calculate ρ_2.

Total system mass m_{total} has decreased by $\dot{m}_{vapour} \Delta time$

Then increment the sim time to a new time 2, and perform the vapour emptying calculation loop again.

Note that towards the end of the vapour-burning phase, the vapour temperature can get seriously chilly: 161 Kelvin. At such low temperatures, it's debatable whether the Nitrous would burn, so I stop the sim when the vapour's pressure drops below 2.5 Bar.

Chapter 4: How a rocket engine works - pressure and flow

In later chapters we'll design a Nitrous oxide hybrid rocket engine to get one person into Space, launched from a gas balloon at 70,000 feet altitude (21.3 kilometres). However, before we can design this rocket, you'll have to understand in-depth how rockets work.

So in this chapter we'll look in detail at a new way of describing how a rocket produces its thrust. Then in the next chapter, I'll introduce rocket science parameters that can be optimised to get the best performance out of the rocket.

Pressure

Let's start with the most basic truth:

Chemical rockets work because of pressure, it's the pressure force pushing on the different parts of the inside of the rocket than makes the thrust. Don't let anyone tell you differently, or try to obscure this truth with excessive mathematics (judge for yourself whether I'm guilty of that!).

[Electric rockets use a potential difference: a field of 'electrical pressure'.]

Nozzle thrust

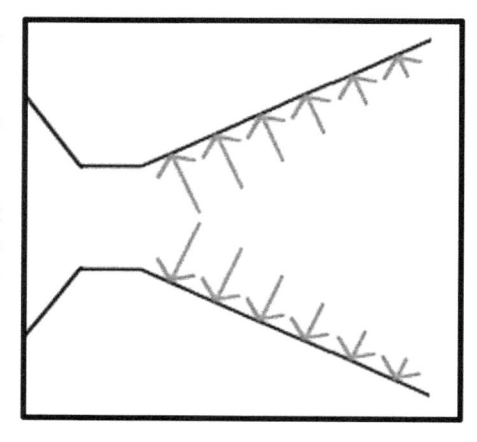

You can get extra thrust from the nozzle. Actually, considerably more thrust, perhaps increasing the overall thrust of the engine by 1.5 times.

The gas flowing through the nozzle is still at a fairly high pressure (admittedly lower pressure towards the end of the nozzle).

Notice that the arrows get shorter (indicating lower pressure) towards the exit of the nozzle, indicating that the pressure drops steadily down the nozzle. The exact variation of the pressure along the nozzle wall is known as a **pressure distribution**.

Now the nozzle wall is curved (it's some sort of cone) and the slope of its wall may change along the nozzle if it's a Bell nozzle (see page 62). So each small section of nozzle wall can point in a different direction, therefore the normal direction that the pressure points in will also be in different directions at different parts of the nozzle.

The only way we can sum the effects of all these different directions together is to sub-divide the nozzle wall into hundreds of tiny elemental areas, deal with each area separately by working out the thrust force on that area, then sum (integrate) these little thrust forces together.

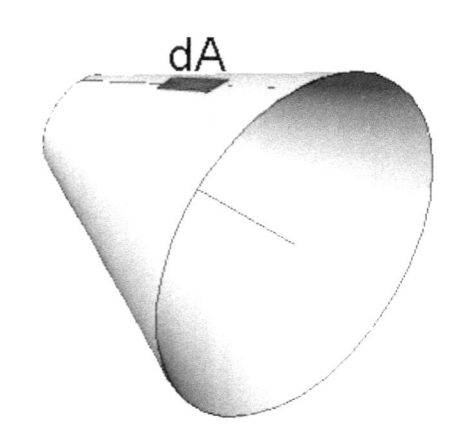

So let's look at an elemental area of nozzle wall near the roof of the nozzle which we'll call dA. It's drawn as a dark rectangle here:

If we add together all the forces from the pressure distribution (shown by the small arrows) that occur across area dA, we get one large force which is equal to that pressure times area dA. We can then resolve this into horizontal (along the nozzle) and vertical (across the nozzle) components by drawing a right-angled triangle to get the thrust from this area.

Now the vertical component is larger than the horizontal component, but it doesn't do anything useful; it just stretches the metal of the nozzle wall. We'll ignore it from now on.

Using trigonometry, the horizontal thrust component is the total force times the sine of the angle of the wall where element dA is located:

$$horizontal\ thrust = total\ force \times \sin(wall\ angle) \quad \text{equ. 4.1}$$

Now the total force is equal to the pressure times area dA so:

$$horizontal\ thrust = pressure \times dA \times \sin(wall\ angle) \quad \text{equ. 4.2}$$

Once we have the thrust from dA, we then do the same for all the other hundreds of dA's that make up the nozzle wall, and sum (integrate) the thrusts together to get a total thrust.

Too much like hard work

So now we know how the rocket makes its thrust. If we can work out the pressure distribution on all the little elemental areas of the internal walls of the combustion chamber and nozzle we can calculate the thrust.

But it will involve a rather painful summation, especially as the walls of the combustion chamber and nozzle tend to be curved rather than straight, which then needs hundreds of little vector diagrams for each little individual bit of curved wall. (We pretend that a curve is made up of dozens of little straight lines joined together.)

We'll end up having to do hundreds or thousands of calculations for every little tiny area dA of wall.

But actually, working out what the pressure distribution will be can turn out to be difficult to do in practice, it requires powerful computers that can calculate how the gas swirls and flows all around the combustion chamber and down the nozzle, because moving gas has a lower pressure.

What we need is a much simpler way of analysing the thrust, and in fact rocketeers of old had no choice as there were no powerful computers back then.

So now I'm going to talk about a different, simpler, way to analyse the thrust. I say 'analyse' rather than 'calculate' because the actual numerical value of the thrust is less important than how efficiently this thrust is produced, which is what we'll concentrate on in the first half of chapter 5. After all, we can easily measure the thrust mechanically on the ground with a force measuring device.

In a way, it's more difficult to picture this new analytical method in your head as it's physically rather far removed from what's actually going on with the pressure. Instead this method works with side-effects and consequences of the pressure rather than directly with the pressure.

But always remember that what I talk about next is *caused* by the pressure.

Newton's 1st and 2nd laws tell us that the consequence of the gas pressure within the combustion chamber is a continuing increase of gas momentum down the nozzle (see page 5). The gas's inertia and the continuing drop in pressure down the nozzle act together in a circular cause-and-effect relationship that establishes a finite gas flow down the nozzle and maintains the pressure difference between inside the chamber and outside. Without 'something to push against' (gas inertia) the gas would exit the chamber instantly, and so the gas pressure in the chamber would plummet instantly, therefore so would the thrust. So the gas inertia is a vital part of the functioning of a rocket.

Let's consider this gas flow through the nozzle:

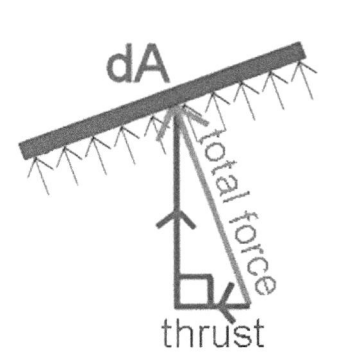

Mass flow rate and exhaust velocity

Recall that the nozzle is a specially shaped pipe bolted to the end of the rocket. Its job is to get the gasses coming out of the end of it up to the highest possible speed. We'll find out why this is important shortly.

To describe the nozzle, (and *only* the nozzle, not the rest of the rocket) consider the flow of two rivers: a swiftly moving stream, and a slow but mighty river.

The stream has a high speed as it rolls down the mountainside but a low flow rate of water mass: it's not carrying much water along with it.

Whereas the mighty river is wide and deep, it's carrying a huge *amount* of water along with it; its flow rate of water mass is huge, thousands of kilograms per second are flowing downstream even though its speed might be quite slow.

Flow rate and speed are *two completely different ways* of describing a flow, although they're not unconnected.

The flow of fluid mass is the mass flow rate. This is measured in kilograms per second, and is given the symbol $\dot{m}$ where the dot above the 'm' was Isaac Newton's fluxion symbology for 'rate with respect to time'.

To recap, I've just introduced to you a new way of thinking about the word 'flow'. When people say 'a fast flowing river' they're linking the word 'flow' to the word 'speed'.

But a rocket engineer wouldn't say this. He'd say 'the river is moving quickly' and then he'd talk about a *completely separate thing*, 'the river's flow rate is quite low' by which he means that the *amount* (in kilograms) of water going past him every second is low.

It's vitally important for rocketeers to understand the difference.

In rocketry, the flow *speed* is called the **exhaust velocity** V_e (measured in metres per second) where 'exhaust' denotes the gas that is exiting the end of the nozzle.

Velocity is the vector of speed: speed where you're concerned with the direction that the speed is going. Exhaust velocity is the speed of the exhaust gas in the direction straight out the nozzle (parallel to the nozzle axis). If the exhaust isn't coming straight out, the speed may be the same, but the thrust will be in some other direction.

But now let's talk about side effects. A side effect of the gas pressure pushing on the inner walls of the combustion chamber and nozzle to create the thrust is that the gas gets pushed out of the nozzle (action and reaction in this case), and this mass flow rate of gas getting pushed out can be measured. Remember that the action force causes the gas to get pushed out, but it's exactly equal to the reaction force. So if you measure the mass flow rate of gas leaving the nozzle, then you're well on your way to being able to calculate the reaction force which is the thrust. This may seem an odd way to calculate the thrust, but bear with me.

A high exhaust velocity is important

Now it so happens that the nozzle shapes that make best use of their nozzle pressure distribution to give the highest nozzle thrust just happen to be the shapes that also give the highest exhaust velocity. *So a high exhaust velocity tells you that your nozzle shape is giving high thrust: it's working efficiently.*

Why?

Let's examine a chunk of exhaust gas - a thick disc of gas - as it travels down the nozzle.

The pressure in the nozzle drops with distance travelled down the nozzle. Therefore all down the nozzle, the chunk of gas (shown by a coloured strip here) has a higher pressure pushing on its upstream end than is pushing on its downstream end. (Remember that within the gas, the pressure pushes in all directions; we're concentrating on the component of the pressures pushing on either side of the chunk of gas here.)

So the larger upstream pressure wins and causes a net force that pushes the gas chunk down the nozzle. As we'll see shortly, the chunk of gas responds to this net force by accelerating to higher speed.

But this only works as long as there's a higher upstream pressure pushing the chunk along.

If the pressure distribution has been badly affected in some way, or if it reaches zero pressure long before reaching the end of the nozzle, then the upstream force disappears and the chunk's acceleration to higher speed falters.

Then its total increase in speed by the time it reaches the nozzle exit (which is the exhaust velocity) is lower than it should be, showing that something's wrong with the pressure distribution.

So, to sum up: thrust is caused as a reaction force to the action force of gas pushing itself away from the rocket. The more gas pushed away every second (the mass flow rate of gas atoms out the nozzle), *and* the faster it's pushed away (the exhaust velocity) indicate a higher action force and so indicate a larger thrust.

The thrust equation

Actually, the mass flow rate and exhaust velocity do more than just indicate the thrust: if you multiply the nozzle's mass flow rate ($\dot{m}_{nozzle}$ in kilograms per second) by the exhaust velocity (V_e in metres per second) the answer is actually *equal* to the rocket's thrust (T), in Newtons:

thrust = nozzle mass flow rate × exhaust velocity

or,

$$T = \dot{m}_{nozzle} V_e \quad \text{equ. 4.3}$$

This is a useful thrust equation, but is of more use if we rearrange it so that we can analyse the exhaust velocity to make sure it's sufficiently large:

$$V_e = \frac{T}{\dot{m}_{nozzle}} \quad \text{equ. 4.4}$$

There are other mass flow rates and velocities inside the rocket, but it's only what's coming out the nozzle (the exhaust) that you multiply together to get the thrust.

So the mass flow rate times the exhaust velocity is *mathematically* equal to the thrust *but they do not cause the thrust*. As we've seen it's the pressure pushing on the inner walls of the combustion chamber and nozzle that causes the thrust force. The mass flow rate and exhaust velocity are just handy *side effects* that we can easily conceptualise to calculate the thrust.

Now, some physicists and rocketeers will argue with me here: they think the nozzle flow *does* cause the thrust. But that's because they've deliberately chosen to work with the acceleration and flow of the combustion chamber gas (using Isaac Newton's 2nd law, see below) rather than work with the pressure force that caused this acceleration of the gas in the first place.

Where does this thrust equation come from? It comes from the original form of Newton's 2nd law as discussed in chapter 1:

$$F = \frac{d}{dt}(mV) \quad \text{equ. 4.5}$$

With a rocket, the nozzle exit velocity (the exhaust velocity) is constant, while mass is continually being lost by the engine, so mass *m* is changing with time:

$$F = \frac{d}{dt}(mV) = V\frac{d}{dt}(m) = V\left(\frac{dm}{dt}\right) = \dot{m}V \quad \text{equ. 4.6}$$

From the Law of conservation of momentum (chapter 1), if mass is continually lost from the engine (plus rocket vehicle), then the engine (plus rocket vehicle) changes its momentum (its velocity) in the thrust direction.

So the quantity $T = \dot{m}_{nozzle}V_e$ is known as the **momentum thrust.**

Why bother?

Why don't rocketeers simply measure the thrust force rather than bothering with mass flow rate and exhaust velocity?

Well you certainly can measure the thrust quite easily, and you can even measure the pressure at the cooler front wall of the combustion chamber.

But when you want to get the most out of your rocket, then you'll want to get as much thrust from your nozzle as well. It's very difficult to measure the pressure within the nozzle because not only is the flow red-hot in there which melts your pressure sensors, but also the pressure is different all through the nozzle (a non-constant pressure distribution). You'd need hundreds of pressure sensors all down the nozzle to get an accurate picture of the changing pressure, and so many sensors would interfere with the nozzle flow and cause a loss of thrust; it's very hard to do.

Instead, you can concentrate on trying to get the highest exhaust velocity out of the nozzle because you now know that this shows that the nozzle is working at its best and so you'll get the most thrust for the least amount of fuel burnt.

As we'll see in chapter 6, we can use energy (the internal energy and pressure-work energy of the hot gas inside the rocket) to quickly and easily calculate the exhaust velocity, and so get the thrust. So it makes sense to deal with mass flow rate and exhaust velocity when using this energy approach, which is the approach that rocket scientists use nearly all of the time.

Having said that, using mass flow rate and exhaust velocity is a bit of a rocketry habit: the German rocketeers in World War Two decided to analyse the thrust this way, and everyone's copied them.

Changing velocity

Let's go back to the river. If the river banks were shaped in a particular way (getting ever closer together) then the speed of the river would increase.

And so it is with the exhaust velocity in a rocket. It starts at near zero speed as the gas enters the nozzle, steadily getting faster as it flows through the nozzle, and comes out the end of the nozzle with a very high speed. We'll find out why shortly.

One thing to realise is that although the speed differs along the nozzle, *the mass flow rate is the same from one end of the nozzle to the other*. This requires some thought, but because no mass was added nor taken away inside the nozzle, the rate that mass flows out of the nozzle can only be equal to the mass flow rate into the nozzle. This is an example of the Law of mass continuity discussed in chapter 1.

So $\dot{m}_{nozzle}$ is the mass flow rate flowing past any point along the nozzle you care to look at.

(There are other mass flow rates inside the rocket, and they can be different to $\dot{m}_{nozzle}$, but we're only talking about the nozzle for now.)

Nozzle mass flow rate

So from the above thrust equation, one way to increase the thrust is to design the rocket to increase the nozzle mass flow rate ($\dot{m}_{nozzle}$).

From the Law of mass continuity, to get a mass flow rate *out* of the chamber and down the nozzle, then you have to make a mass flow rate *into* the chamber.

This flow rate in comes from an inflow of propellants.

So:

$$\dot{m}_{in} = \dot{m}_{fuel} + \dot{m}_{oxidiser} \qquad \text{equ. 4.7}$$

and:

$$\dot{m}_{out} = \dot{m}_{nozzle} \qquad \text{equ. 4.8}$$

You can make the nozzle mass flow rate as large as you like, but a very high number of kilos per second pouring through the nozzle means that all the kilos in the rocket vehicle get used up very quickly. Those kilos come from the propellants, so the propellant tanks empty quickly.

Therefore you end up with a very short but violent thrust, more of a bang than a whoosh. Such a large thrust will break the rest of the spacecraft (and you too if you're sat atop it).

To tailor the thrust to just the right amount that you require, you go for maximum exhaust velocity, then adjust the nozzle mass flow rate until the thrust is just right, and not too high. (Remember you multiply nozzle mass flow rate and exhaust velocity together to get the thrust.)

To do the same thing, you could in theory go for maximum nozzle mass flow rate then adjust the exhaust velocity, but for rockets this has been found to be inefficient. [However certain jet engines called turbofans do indeed do it this other way round.]

Humans can't take high gees so what you want for a human-carrying rocket is a long, low thrust stretched out over a longer time, so you want a lowish nozzle mass flow rate.

To get a lowish mass flow rate *out* of the combustion chamber and down the nozzle simply make a lowish mass flow rate of propellants *into* the chamber.

[With hybrid engines, the mass flow rate 'into' the chamber is how many kilograms of its plastic fuel are turned into a gas every second ($\dot{m}_{fuel}$), plus the mass flow rate of oxidiser coming into the chamber.]

We always turn the propellants into a gas within the combustion chamber (in some rockets the propellants may already be gaseous). The mass flow rates don't care whether they're liquid or gas, but we'll discover shortly that we need gas in order to get a high exhaust velocity.

So, we've tailored the nozzle mass flow rate to the exact amount we need simply by choosing the right mass flow rate of propellants in. But now we're going to do something that might seem odd at first, we're going to dam-up the combustion chamber by only letting the flow pass out through a small hole.

The smaller the cross-sectional area across the narrowest part of the nozzle (called the throat), the less mass can flow through it every second.

When rocketeers denote the nozzle throat, they usually give it a star (*). So the throat cross-sectional area is written as A^*. The pressure at the throat is P^*, and the gas's temperature at the throat is T^*, and so on.

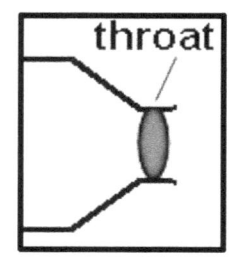

The nozzle mass flow rate changes directly with A^*. If you halve A^* you halve the nozzle mass flow rate that passes through it.

Why does this happen?

Here's a diagram showing the pressure forces within the combustion chamber. Notice that I've only drawn the pressure (between the dotted lines) that isn't balanced or cancelled-out: the pressure on the top wall of the chamber is cancelled out by the pressure on the bottom wall for example.

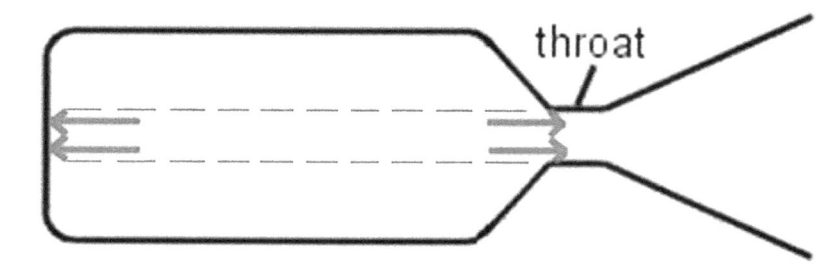

But notice that most of the pressure on the forward wall of the chamber is balanced by the pressure on the back wall of the chamber. Only between the dotted lines where there's a hole (the throat) is there an unbalanced pressure available.

This unbalanced pressure acting on the area between the dotted lines causes two things to happen:

1) It pushes on the front wall - the action area from chapter 1 - to cause the thrust force (the action).

2) It is also the force that squirts the mass out the nozzle throat: it causes the nozzle mass flow rate (the reaction).

Note that the force caused by the unbalanced pressure is equal to the pressure times the throat cross-sectional area: the smaller the throat area, the less force there is to cause the thrust force, and also to push the mass flow rate through, so they both decrease by the same amount.

So that's why - in the thrust equation - the thrust depends on the nozzle mass flow rate: it's *actually* depending on the area between the lines on the front wall, (the action area) which equals the throat cross-sectional area that the nozzle mass flow rate depends upon.

Because the nozzle mass flow rate is the same all through the nozzle, it's the throat that sets the flow rate all through the nozzle.

Nozzle mass flow rate and throat temperature

Another thing that affects the nozzle mass flow rate is temperature.

Cold gas flow passes through the throat more easily than hot gas flow. This is because as a gas's temperature rises, the more vigorously and more often the gas atoms bounce off each other.

This makes the atoms spread farther apart (**expansion**: a drop in density) the higher the temperature gets.

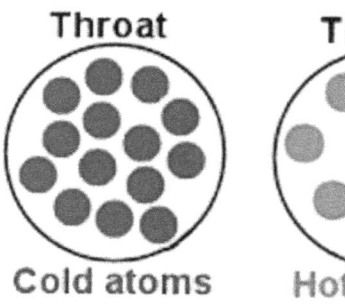

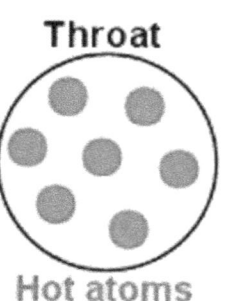

Because they're further apart this means that fewer of them will fit down the hole of the throat every second.

So if the flow is hot (large T^*), as it is in a rocket, then you have to make the throat a bit bigger (larger A^*) to get the nozzle mass flow rate you want. This turns out to be a *very* important fact, so do remember it!

How much bigger do you have to make the throat?

The above simple explanation implicitly assumes that the hot and cold flows pass through the nozzle throat at the same speed, but actually they don't: the hot flow passes through faster. Does this destroy the explanation? Let's investigate mathematically:

The Law of mass continuity applied to the flow through the nozzle throat can be described by the mass continuity equation (re-arranging equation 1.18):

$$A^* = \frac{\dot{m}_{nozzle}}{\rho^* V^*} \quad \text{equ. 4.9}$$

where ρ^* and V^* are the gas density and velocity at the throat.

[The mass continuity equation doesn't just work for nozzle flow, it works for all flows of liquids or gasses along pipes, and also for flow through the holes in the rocket's injector which is where we'll use it in chapter 9. Also note that the mass continuity equation shows that mass flow rate and flow velocity are coupled together by flow density.]

Now using the Ideal gas equation (equation 1.14) at the throat:

$$\rho^* = \frac{P^*}{R T^*} \quad \text{equ. 4.10}$$

And we'll discover later that at the throat, the gas velocity is at Mach 1, the speed of sound a, where

$$a = \sqrt{\gamma R T} \quad \text{equ. 4.11}$$

(where T is the gas temperature, in Kelvin of course).

γ is the 'isentropic exponent' also called 'the ratio of specific heats' of the gas (equation 1.22).

so:

$$V^* = a = \sqrt{\gamma R T^*} \quad \text{equ. 4.12}$$

Therefore combining the mass continuity equation 4.9 and the Ideal gas equation:

$$A^* = \frac{\dot{m}_{nozzle}}{\rho^* V^*} = \frac{\dot{m}_{nozzle}}{\left(\frac{P^*}{R T^*}\right)\sqrt{\gamma R T^*}} = \sqrt{T^*}\left(\frac{\dot{m}_{nozzle}\sqrt{R}}{P^*\sqrt{\gamma}}\right) \quad \text{equ. 4.13}$$

Take a look at what has happened here: in the denominator of the second equation of 4.13, we see that although the hot flow is passing through the nozzle faster, it's only faster by the *square root* of temperature. Whereas the drop in flow density depends directly on temperature, so is the larger effect that wins out. So our original statement holds true: A^* has to increase with (the square root of) the gas temperature at the throat.

Keep this in mind: we'll discover shortly that lots of the things that affect the rocket's thrust *depend on the square root of the gas temperature.*

Mass flow rate and combustion chamber pressure

The purpose of the narrow nozzle throat is to stop the gas in the chamber leaking away until it can build up to high chamber pressure (we *need* high chamber pressure to get a high pressure on the action area, and a high pressure distribution on the nozzle).

Let's look at this more closely:

We've got our required mass flow rate of propellants into the chamber, but now we've blocked up the way out, leaving a small hole that'll only let a small dribble of mass flow rate out.

So now the mass flow rate coming into the chamber and getting turned into gas ($\dot{m}_{in}$) is greater than what can flow out of the chamber and out of the nozzle ($\dot{m}_{out}$).

What happens then? There's too much gas heading for the nozzle that can't get out, where does it go? From a non-steady-with-time form of mass continuity equation, it can't go anywhere, it just builds up and up inside the combustion chamber just as water will pile up behind a leaky dam across a river. The chamber gas mass ($m_{chamber}$) increases with time:

$$\frac{d}{dt}(m_{chamber}) = \dot{m}_{in} - \dot{m}_{out} \quad \text{equ. 4. 14}$$

As more and more gas atoms fill the chamber, the more often they bang into one another and the walls. So the force on the walls increases: the pressure inside the chamber goes up.

From the above equation, and the Ideal gas equation 1.14:

$$P = \rho R T \quad \text{equ. 4. 15}$$

and the density equation:

$$\rho = \frac{m}{Vol} \quad \text{equ. 4. 16}$$

where *Vol* is the fixed volume of the combustion chamber, we can describe the rate of increase of the combustion chamber pressure with time:

$$\frac{dP}{dt} = RT\frac{d\rho}{dt} = RT\frac{1}{Vol}\left(\frac{d}{dt}(m_{chamber})\right) = \frac{RT}{Vol}(\dot{m}_{in} - \dot{m}_{out}) \quad \text{equ. 4. 17}$$

As the pressure inside the chamber gets larger, this pushes the nozzle mass flow rate out of the chamber with a larger force and so the nozzle flow rate out gets larger. Eventually, the outflow gets large enough that what flows out exactly equals what flows in again ($\dot{m}_{out} = \dot{m}_{in}$) and the chamber pressure rises no more.

This all happens within a second or so of the engine being lit, you can actually hear a rocket building up chamber pressure as a whooshing/whistling noise just after ignition, before the roaring noise begins.

The trick with designing rocket engines is to carefully choose the nozzle throat size (area A^*): make A^* small as this'll build up the chamber pressure nice and high. But if the chamber pressure is too high the chamber walls will burst so choose A^* with care.

[Note that equation 4.17 is strictly for a liquid engine only, where combustion chamber volume '*Vol*' doesn't change with time. But for a solid or a hybrid, this volume steadily (but slowly) increases as the fuel port widens. We deal with this extra issue in chapter 9]

The combustion chamber pressure is usually very high, so is measured in either Bar, or MegaPascals (MPa). One Bar is ten times smaller than a MPa so one MPa = 10 Bar.

Nozzle mass flow rate and choked flow

You saw earlier in this chapter that it's the combustion chamber pressure ($P_{chamber}$) that pushes the gas through the throat.

Now when a gas pushes on a wall, it creates a force. Only the pressure at the wall matters. But when pressure is pushing a flow of liquid or gas along, the flow is caused by a *difference* in pressure (ΔP = higher minus lower) between one end of, say, a pipe, and the other.

So when pressure causes a flow, there are *two* pressures involved, the higher upstream pressure and the lower downstream pressure. We saw this on the diagram on page 47 where two pressures were pushing on either side of a chunk of nozzle gas: the higher pressure speeds the flow up in the direction of its push while the lower pressure pushes back against the higher pressure so slows the flow down a bit.

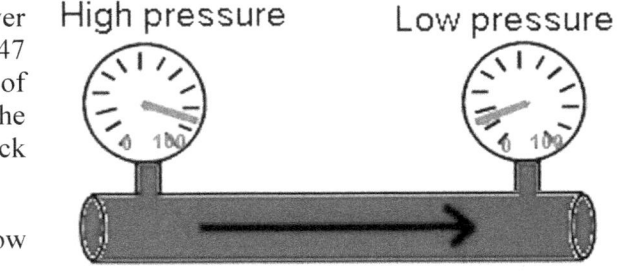

Here's a pipe with two pressure gauges showing the flow flowing from the high pressure end to the low pressure end:

The pressure in the combustion chamber is high, while the pressure at the nozzle exit is pretty much the pressure of the atmosphere outside the rocket, which is low.

So you'd think that the bigger the difference in pressure between the combustion chamber and the nozzle exit

($\Delta P = P_{chamber} - P_e$), then the bigger would be the nozzle mass flow rate.

With almost all types of flow you'd be absolutely right, and the exhaust velocity does indeed depend on this ΔP.

And as the combustion chamber begins to build up pressure just after ignition you'd still be right, the nozzle exit pressure matters to the nozzle mass flow rate. But once the combustion chamber reaches a certain pressure higher than the nozzle exit pressure, something very odd happens to the nozzle mass flow rate alone:

As long as this pressure difference is larger than about 2 Bar, *the nozzle mass flow rate depends only on the chamber pressure*, the pressure at the nozzle exit now has no effect.

If the chamber pressure was 50 bar, the nozzle exit pressure could be 45 bar or 0 bar, and the nozzle mass flow rate wouldn't change.

This strange effect is called **choking**, the nozzle is said to be choked at the throat, and it's caused by the fact that the nozzle gas flow always hits Mach 1 at the throat, whatever the throat size or shape.

Now pressure travels at the speed of sound through a gas, no faster. If the throat is *at* Mach 1 (the speed of sound) then it's impossible for any downstream nozzle pressure changes to flow fast enough to make it upstream of the throat.

Remember that pressure is caused by the gas molecules zipping around at high speed and colliding with each other. Though some molecules move faster than others, they have a sort of overall average speed, and so pressure can't travel through the gas faster than this average speed. [The speed of sound is about 7/10ths of the mean speed of the gas molecules.]

So everything downstream of the throat is 'cut off' from upstream of the throat as far as pressure is concerned: upstream simply can't sense the downstream pressure.

Now when I say that the nozzle mass flow rate depends only on the combustion chamber pressure, that isn't strictly correct because we know that flows depend upon two pressures, an upstream pressure and a downstream pressure.

Therefore the size of the mass flow rate of a choked nozzle is set up by the difference in pressure between the pressure at the end of the combustion chamber just as the gas flows into the nozzle, and the pressure at the throat.

However, for a combustion chamber that isn't excessively narrow, this pressure difference depends *only* on the combustion chamber pressure because this pressure difference is *always the same fraction* of the chamber pressure. So in the rocketry equations, only the chamber pressure needs to be given which is why we can say that only the chamber pressure matters.

Neither of these two pressures (chamber or throat) can know about the pressure downstream of the throat if the nozzle is choked. The mass flow rate *downstream* of the throat is then maintained purely by the molecules pushing on one another as they flow down the nozzle; mass continuity, nothing to do with the upstream pressure.

[There is a caveat to this, which is of course that the molecules don't roll along against one another like marbles down the nozzle; as we've seen they collide with each other, so mass continuity is actually maintained by local pressure forces. *But* these are pressure forces *which move downstream at the flow speed*. Downstream of the fixed throat, very few inter-molecule collisions have enough velocity in an upstream direction to effect a pressure change heading upstream relative to the nozzle throat.]

The exhaust velocity on the other hand responds to the changing pressure all along the nozzle from upstream of the throat to the nozzle exit. It responds to the dropping local pressure - carried along at the flow speed - by expanding, which accelerates the flow speed as I'll describe shortly.

The nozzle mass flow rate changes directly with combustion chamber pressure: double the chamber pressure and you double the nozzle mass flow rate because you've doubled the force caused by the pressure that is pushing the flow along.

Propellant tank pressure

For the propellants to flow into the combustion chamber, then the beginning of the pipe they flow in on has to be at a higher pressure than the chamber pressure.

In this book, I've chosen to use Nitrous oxide because it has a very high vapour pressure (see chapter 2). Just as with a water/aquajet rocket, this vapour pressure is used to force the liquid Nitrous out of the tank and into the combustion chamber.

How much of the Nitrous in the vehicle tank is in the liquid state simply depends on how much liquid you filled the tank with.

Measuring the nozzle mass flow rate

You can't measure the nozzle mass flow rate if the rocket has a hot exhaust because any measuring devices you put in the scorching hot exhaust flow will quickly melt.

But as I explained earlier, the chamber pressure keeps rising until the mass flow rate *out* of the chamber (the nozzle mass flow rate) equals the mass flow rate *into* the chamber (the mass flow rates of the propellants). So $\dot{m}_{out} = \dot{m}_{in}$

We can easily measure the flow rates of the propellants 'into' the chamber as they're cold, and we now know that they'll be equal to the nozzle flow rate out over most of the burn, except for the first second or so after ignition.

A simple way to measure the mass flow rates 'in' to our hybrid is to weigh both the Nitrous tank and the solid fuel grain before and after the burn, and also time how long the burn took.

If 40 kilos of Nitrous were emptied from the tank in a 10 second burn, then on average, the average mass flow rate of the Nitrous oxidiser into the chamber was:

$$\dot{m}_{ox} = \frac{40}{10} = 4\ kilos\ per\ second \quad \text{equ. 4. 18}$$

Similarly, if 20 kilos of plastic fuel were burnt in 10 seconds, then the average mass flow rate of the fuel 'in' was:

$$\dot{m}_{fuel} = \frac{20}{10} = 2\ kilos\ per\ second \quad \text{equ. 4. 19}$$

So the average nozzle mass flow rate $\dot{m}_{nozzle}$ would have been:

$$\dot{m}_{nozzle} = \dot{m}_{ox} + \dot{m}_{fuel} = 4 + 2 = 6\ kilos\ per\ second \quad \text{equ. 4. 20}$$

There are turbine devices that can measure propellant flow rate, but they can fail catastrophically in Nitrous. The turbine causes the Nitrous to vaporize, which turns it into a jet engine: the turbine's revolutions increase to destruction.

There is another type of flow meter called a Coriolis flow meter that measures the 'centrifugal force' of the Nitrous speeding round a curve in a pipe. However, you need a very accurate measure of the Nitrous density as it flows round the curve, which can be hard to deduce.

Exhaust velocity and the nozzle shape

So we now have a nozzle with a small hole within it so that the flow out is dammed-up and we get a combustion chamber full of high pressure gas.

The reasons for doing this are to get a large pressure force on the front wall of the combustion chamber to get a large chamber thrust, but equally importantly to get the high pressure distribution pushing on the inside of the nozzle downstream of the throat to get a high nozzle thrust (see page 44).

As a side effect that we can measure to make sure the nozzle's working well, we'll get a very high exhaust velocity.

Let's discover how this works:

From the mass continuity equation 1.18, the nozzle mass flow rate doesn't care what shape the nozzle is, it only cares how wide (the cross-sectional area A^*) the nozzle throat is:

$$\dot{m}_{nozzle} = \rho\ A^*\ V \quad \text{equ. 4. 21}$$

The exhaust velocity on the other hand *does* care what shape the nozzle is. We can carefully shape the nozzle in such a way as to get the maximum exhaust velocity out of the pressure available, which shows us that the pressure distribution within the nozzle is being used efficiently to create thrust.

Think of a water tap. There's a certain pressure pushing the water into your house just as the chamber pressure pushes the exhaust gasses into the nozzle.

If you just turn the tap on, the water runs out slowly.

But if you put your thumb over the tap to make the hole narrower you can get a much faster water velocity, it sprays out at high speed.

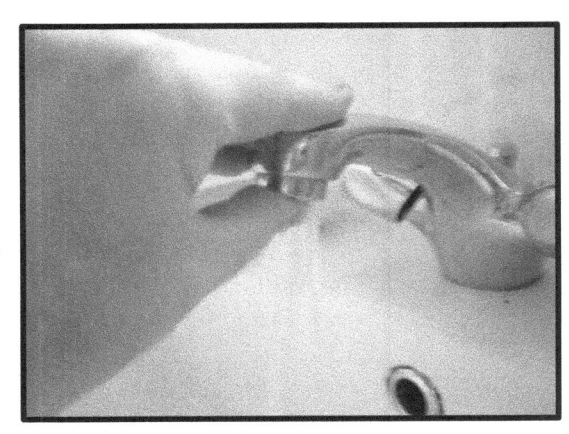

So rocket engineers in the early days put a pipe on the end of the combustion chamber that got progressively narrower.

A pipe designed to speed up the flow is called a nozzle.

What's happening to the flow inside this shape of nozzle?

It's simply following the Law of mass continuity as described by re-arranging the mass continuity equation 1.18:

$$V = \frac{\dot{m}_{nozzle}}{\rho A} \quad \text{equ. 4.22}$$

And - this is very important - keeping the gas density ρ constant.

This equation shows that as A decreases, V increases.

Nozzle shape: de Laval's nozzle

But as the early rocketeers found out, there was a problem with this shape of nozzle. The exhaust gas would hit a speed-limit of a few hundred miles an hour as it left the end of the nozzle (quite fast) but wouldn't go any faster.

They tried making the combustion chamber pressure higher to give as high a pressure difference down the nozzle as possible to turn into speed, but the gas stubbornly refused to go faster which showed that the extra pressure wasn't causing extra thrust.

An engineer called Gustaf de Laval solved the puzzle. He realised that the gas had reached a speed-limit which was Mach 1, the speed that sound travels through the gas.

Gas moving slower than the speed of sound (called **subsonic**) behaves very differently to gas moving faster than the speed of sound (called **supersonic**).

At subsonic speeds, a *narrowing* of the pipe speeds up the gas. But with supersonic flow, this gets turned on its head, and a *widening* of the pipe speeds up the gas.

It seems counterintuitive, but it's true, I'll explain how this happens shortly.

De Laval built a nozzle that got progressively narrower while the gas was subsonic, but as soon as the gas went supersonic, he made his nozzle progressively wider.

And it worked, his gas exited the end of his nozzle at very high exhaust velocity, around Mach 4 (four times the speed of sound).

So rocket nozzles are built the de Laval way to get a high supersonic exhaust velocity, which we know means that they're using the pressure efficiently to create nozzle thrust.

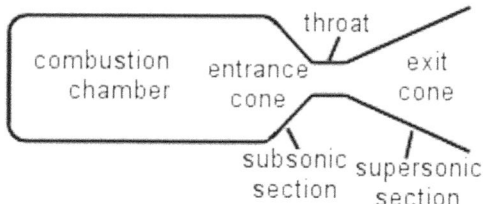

Interestingly, no matter how wide or narrow the nozzle throat is, the gas always adjusts itself to hit exactly Mach 1 at the throat ($M^* = 1$).

What happens is that as the rocket builds up combustion chamber pressure just after ignition, the flow speed is at its fastest at the narrow throat (from the mass continuity equation).

56

As the chamber pressure rises, the flow speed at the throat gets ever faster until it hits Mach 1. The physics of the flow then force the portion of the flow doing Mach 1 always to remain at the throat as the chamber pressure rises further. This is because the fundamental behaviour of the flow changes when it goes from below Mach 1 to above Mach 1, as I'll explain below when I talk about expanding gas.

Then the flow goes supersonic downstream of the throat.

De Laval nozzle: expanding gas

How exactly does the gas flowing down a de Laval nozzle reach high speed?

The pressure in the combustion chamber is higher than at the nozzle exit, so there's a force pushing the gas through the nozzle at quite a speed. But there's something else going on.

I know this is a very unlikely scenario, but if you happen to be at the bottom of a deep swimming pool, and you happen to have a balloon with you, blow it up a little, tie it closed, then let it go.

As it floats upward, the water pressure around it lessens. (The water pressure increases the deeper you go because of gravity.) So as the balloon rises there's less pressure squashing it. So the gas in the balloon can expand (swell up), the balloon gets bigger the higher it rises.

In the same way, the gas expands (*its density drops*) as it passes down the nozzle because the pressure drops with increasing distance down the nozzle. (It starts at the chamber pressure and steadily drops down to the lower pressure at the nozzle exit.)

The mass flow rate still remains the same all down the nozzle, so of course the mass continuity equation still applies, although *now* we've got to add in the effect of changing density ρ:

$$V = \frac{\dot{m}_{nozzle}}{\rho A} \quad \text{equ. 4.23}$$

Now earlier we saw how the gas would speed up in the narrowing subsonic section, purely by the effect of decreasing cross-sectional area A (without expanding, therefore keeping ρ unchanging in this equation). Well, surely the gas would then slow down in the supersonic section which gets steadily wider?

It *would* slow down if it wasn't expanding, but it *is* expanding.

There are two competing processes going on: in the subsonic converging section of the nozzle, there is almost no expansion occurring (the density hardly changes), so the flow behaves like water. Only the narrowing of the nozzle speeds the flow up.

But in the supersonic diverging section, considerable expansion occurs (the density drops). The expansion speeds the flow up more than the nozzle shape-changing slows it down. The expansion wins, and so the flow speeds up.

That's the difference with a supersonic gas: expansion wins out, whereas with a subsonic gas, shape-changing wins out. This difference explains why de Laval's nozzle works. Its shape has to change from narrowing to widening once the flow changes from subsonic to supersonic because an expanding flow prefers a widening nozzle for speeding up the flow. A widening nozzle gives the gas room to expand.

Sure enough, the mass continuity equation can describe exactly what's going on downstream of the throat. For easy counting, we'll assume the nozzle mass flow rate is one kilogram per second, so the mass continuity equation (rearranged as above) is:

$$V = \frac{1}{\rho A} \quad \text{equ. 4.24}$$

We'll look at a point just downstream of the throat where the area is 4, and then we'll look at a point further down the nozzle where the area has widened to 8 (twice as wide).

Make ρ equal to 2, and A equal to 4, this gives $V = 0.125$

As the nozzle gets wider downstream of the throat, let's double the size of A to get $A = 8$. But we know that the gas expansion wins, so don't just halve ρ, divide it by 4 to get 0.5

Do the sum and you'll find that V doubles in speed to 0.25, the expansion (the drop in density) wins and makes the exhaust velocity larger.

Remember that the higher the exhaust velocity, the more efficiently the pressure is being turned into thrust. Recall from the diagram on page 45 that it's the horizontal pressure component pushing on the sloping inner wall of the nozzle's exit cone (the part of the nozzle downstream of the throat) that causes the nozzle thrust (the reaction to which speeds the flow up). So de Laval's nozzle works well *purely because* it has a widening exit cone.

Could we not bother with the de Laval shape and just fix a large exit cone right onto the combustion chamber with no entrance cone or throat?

You could, but it wouldn't work. There'd be no throat for the gas to reach Mach 1 at, so the flow would never reach the speed above which it can seriously expand (above Mach 1). [Gas can expand a bit below Mach one, but if the expansion is large (a big density change) it has to occur faster than Mach 1.] Instead, the subsonic flow wouldn't expand. It'd just separate from the exit cone wall and form a narrow jet of gas.

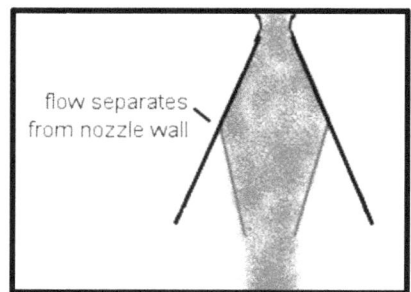

This would cause a region of stagnant air (no speed, low pressure) to push on the exit cone so the thrust from the cone would be feeble.

You can keep the de Laval nozzle fairly simple yet still get a good efficiency: one can make small nozzles as two cones joined together by a short straight parallel throat that's as long as its diameter, as this is easy to make:

A nozzle with a simple straight exit cone is called a **conical nozzle**.

Nozzle area ratio

How much wider the nozzle exit gets compared to the throat determines what the exhaust gas's Mach number will be once it reaches the nozzle exit.

Actually, again it's the cross-sectional areas that count (shown in dark colour here):

The Mach number of the gas at the nozzle exit depends upon:

$$\frac{area\ across\ the\ exit\ A_e}{area\ across\ the\ throat\ A^*} \quad \text{equ. 4.25}$$

This area fraction is called the **area ratio** (or **expansion ratio**) of the nozzle. The bigger the area ratio, the higher the Mach number will be at the nozzle exit so the higher the exhaust velocity. Sometimes the area ratio is given the symbol $\in$. See appendix 3 for an equation to calculate $\in$.

At the throat the speed is Mach 1. At the point down the nozzle where the nozzle is 1.4 times wider than the throat the speed will be about Mach 2 (2 times the speed of sound). And at the point where it is 3.3 times wider than the throat the speed will be about Mach 4 (4 times the speed of sound). And so on. (The exact area ratios depend upon what the exhaust gas is composed of.)

Measuring exhaust velocity

You can't easily measure the exhaust velocity, so what you do instead is to measure the nozzle mass flow rate and the thrust. (To measure the thrust, you fix the rocket to the ground (see chapter 12) and measure the thrust using either a very large spring balance or a **load cell**, which is a device that converts force into an electrical signal.)

You know that:

thrust = nozzle mass flow rate × exhaust velocity or $T = \dot{m}_{nozzle} V_e$ equ. 4. 26

We can rearrange this equation to get:

$$\textit{exhaust velocity} = \frac{thrust}{nozzle\ mass\ flow\ rate} \quad \text{or:} \quad V_e = \frac{T}{\dot{m}_{nozzle}} \quad \text{equ. 4. 27}$$

So knowing the thrust and the nozzle mass flow rate, we can use this equation to work out the exhaust velocity.

The exhaust plume

Another way to gauge the exhaust velocity is to look at the shock diamonds. These are diamond-shaped luminous regions in the exhaust plume of a slightly improperly expanded exhaust, and are caused by shockwaves bringing the exhaust pressure back to ambient.

There's a rule of thumb that the number of diamonds is related to the nozzle exit Mach number: count the number of complete diamonds and subtract one. That's the nozzle exit Mach number (to the nearest whole Mach).

Also, the shape of the exhaust plume tells you if you're using the correct nozzle area ratio. The exhaust should be cylindrical. If it tapers-in, the nozzle flow is over-expanded, or if it bulges out the flow is under-expanded. (See chapter 5 for the definitions of 'over' or 'under' expanded).

To capture shock diamonds, you need to put an infrared filter in front of the camera lens, because digital cameras are over-sensitive to infrared and all they see is a red/yellow solid bar of flame. Having said that, HDPE burning in Nitrous is particularly clean: soot-free, so the exhaust is nearly invisible to the naked eye (it's soot that makes rocket exhausts visible, because soot emits a yellow colour).

More pressure

So both the nozzle mass flow rate and the exhaust velocity increase with higher chamber pressure, and so the thrust does too when you multiply them together. Modern liquid rockets have huge chamber pressures, 150 Bar or more.

As the rocket climbs up through the atmosphere, the pressure of the atmosphere steadily reduces, so the nozzle exit gas can expand more if it's given a chance to. So adding a little extra nozzle is useful here.

In summary

Okay, now we know how to create a high combustion chamber pressure by using a narrow throat, to give us a high exhaust velocity to multiply with our selected nozzle mass flow rate. In the next chapter, we'll analyse this thrust equation to aim for high performance from the rocket.

Chapter 5: How a rocket engine works - performance and heating

In this chapter, we'll build upon the previous chapter, looking at rocketry parameters that characterise the rocket's performance, and why we increase the rocket's performance by heating the gas.

We saw at the end of the previous chapter that the speed the gas leaves the nozzle exit is determined by the nozzle area ratio. The equation that describes this is derived using enthalpy (see chapter 1):

Nozzle exit velocity and specific enthalpy h

In the combustion chamber, the gas as a whole macroscopic lump is moving very slowly, so its speed squared is very low, which makes its kinetic energy practically zero.

Then within the nozzle, the enthalpy energy (see chapter 1) gets converted into kinetic energy, which is the energy of the exhaust velocity (squared) as the gas leaves the end of the nozzle.

How does this happen? Remember that enthalpy includes 'pressure work', the work done by the pressure in moving the gas through the nozzle. (Work in a gas can be expressed as the pressure times the change in the volume of the gas as it expands down the nozzle.)

Using enthalpy gives another way of calculating the exhaust velocity if you know the enthalpy in the combustion chamber (which we'll work out in the next chapter). We'll use specific enthalpy and specific kinetic energy. From the principle of the conservation of energy, the sum of specific enthalpy and specific kinetic energy is constant down the nozzle as we saw in chapter 1, so:

$$h_{chamber} + \frac{V_{chamber}^2}{2} = h_{exit} + \frac{V_{exit}^2}{2} = constant \quad \text{equ. 5.1}$$

where nozzle exit velocity V_{exit} is exhaust velocity V_e

As the gas velocity $V_{chamber}$ in the chamber is almost zero (stagnant), then the specific kinetic energy in the chamber is also practically zero,

so the $\frac{V_{chamber}^2}{2}$ kinetic energy term equals zero.

so:

$$h_{chamber} = h_{exit} + \frac{V_e^2}{2} \quad \text{equ. 5.2}$$

Now enthalpy depends on the gas temperature: $h = C_p T$ (see Appendix 2 for values of specific gas constant C_p)

so:

$$C_p T_{chamber} = C_p T_{exit} + \frac{V_e^2}{2} \quad \text{equ. 5.3}$$

and rearranging this equation gives the exhaust velocity:

$$V_e = \sqrt{2 C_p (T_{chamber} - T_{exit})} \quad \text{equ. 5.4}$$

where you can get T_{exit} from the isentropic flow relations (see Appendix 3).

Appendix 3 can expand this equation (see that appendix) to get the Mach number of the flow at the nozzle exit:

$$M_{exit} = \sqrt{\left(\frac{2}{\gamma-1}\right)\left[\left(\frac{P_{chamber}}{P_{exit}}\right)^{\frac{\gamma-1}{\gamma}} - 1\right]} \qquad \text{equ. 5.5}$$

Now, nozzle area ratio ∈ depends upon this nozzle exit Mach number *and vice-versa*. From appendix 3:

$$\in = \frac{A_{exit}}{A^*} = \left(\frac{1}{M_{exit}}\right)\sqrt{\left[\left(\frac{2}{\gamma+1}\right)\left(1 + \left(\frac{\gamma-1}{2}\right)M_{exit}^2\right)\right]^{\frac{\gamma+1}{\gamma-1}}} \qquad \text{equ. 5.6}$$

So you might think that you ought to make the nozzle exit as wide as possible to expand the gas down to the lowest nozzle exit pressure and hence the highest possible Mach number (highest exhaust velocity).

In Space where the pressure outside the nozzle is zero this is true, you can keep on extending and widening the nozzle as much as you like to increase the Mach number. Though after a while your nozzle will get enormous and therefore heavy, so most rocketeers stop adding extra nozzle sometime before the area ratio equals 100. (70 is a typical maximum for upper stage engines.)

Here's the large area ratio Space nozzle on the back of the Apollo service module above the Moon:

Notice that the nozzle exit is curved rather than being a simple straight cone. This is called a **bell nozzle**, because it looks somewhat like a traditional church bell.

To achieve the same area ratio, a bell nozzle is shorter in length than a simple conical nozzle, which saves mass.

A bell nozzle also changes the pressure distribution pushing on its inner wall into thrust a bit more efficiently.

Why? Because where the pressure is high (just downstream of the throat) the bell nozzle wall is much more angled; larger components of the pressure and wall area are pushing along the thrust direction. But where the pressure is low (near the nozzle exit) then the wall angle is very small.

To work properly, the shape of a bell nozzle has to be designed on a computer using a process called **the method of characteristics** although a skewed parabola is a very good approximation as detailed in appendix 6.

Pressure thrust and effective exhaust velocity

By ceasing to add ever more exit cone before the nozzle gets too large and heavy means that there is still some pressure in the gas as it leaves the end of the nozzle. It would be nice to expand the gas right down to zero pressure to get the best possible thrust, but theoretically, you'd need a nozzle exit infinitely wide!

Still, this little bit of left-over pressure isn't completely wasted, it gives a useful bit of thrust (though not as much as if the nozzle *did* get the gas down to zero pressure.) This is called the **pressure thrust**.

Remember that pressure is a force spread over some area, so to get the pressure thrust's force, simply multiply the remaining pressure by however much nozzle exit cross-sectional area there is:

extra pressure thrust = nozzle exit pressure P_e × area across the nozzle exit A_e

So the total thrust is:

$$T = \dot{m}_{nozzle} V_e + P_e A_e \quad \text{equ. 5.7}$$

The reason we use the nozzle exit cross-sectional area to calculate the pressure thrust is because if we take the horizontal components of elemental areas of nozzle wall (the wall's shadow projected onto a screen oriented in a direction across the throat), we get an area equal to the nozzle exit cross-sectional area *plus* we also include the wall area across the throat. As we know, the area across the throat is really a projection of that part of the area on the front of the combustion chamber wall (the action area) that is equal in size to the throat area.

Effective exhaust velocity

It's a pity that we've had to add this extra pressure thrust term to the end of the thrust equation, it makes the maths a little less clean. Wouldn't it be nicer if we could make this extra pressure thrust term disappear? This isn't trite, some rocketry equations can get seriously long, and reducing this term makes analysis easier.

What we do is to work out the magnitude of the pressure thrust, then convert it so that it becomes an extra exhaust velocity to be added to the actual exhaust velocity. This way, when we multiply this fudged exhaust velocity by the nozzle mass flow rate as usual, we get the same value of thrust:

$$T = \dot{m}_{nozzle} V_e + P_e A_e = \dot{m}_{nozzle}(V_e + extraV_e) = \dot{m}_{nozzle}\left(V_e + \frac{P_e A_e}{\dot{m}_{nozzle}}\right)$$

equ. 5.8

The exhaust velocity that we've cheated and made larger (= $V_e + extra\, V_e$) is called the **effective exhaust velocity**, so-called because it has the same effect as a larger exhaust velocity. It is given the symbol C_e

So:

$$C_e = V_e + extraV_e \quad \text{equ. 5.9}$$

and:

$$T = \dot{m}_{nozzle} C_e \quad \text{equ. 5.10}$$

As a numerical example, suppose a small Space rocket has a total thrust of 10 Newtons, which includes 2 Newtons of pressure thrust. (Pressure thrust is usually a small part of the total thrust.)

If the nozzle mass flow rate is 2 kilograms per second and the actual exhaust velocity is 4 metres per second then the thrust is:

$$T = \dot{m}_{nozzle} V_e + pressure\ thrust = 2 \times 4 + 2 = 10 \quad \text{equ. 5.11}$$

What we have to do now is to keep the first 2 (the nozzle mass flow rate), but then replace the real 4 and the real last 2 by the fake Ce, so that the answer still comes out at 10 Newtons:

$$T = \dot{m}_{nozzle} C_e = 2 \times 5 = 10 \quad \text{equ. 5.12}$$

So the effective exhaust velocity has to be 5 metres per second to keep the thrust at 10 Newtons.

Low altitude nozzles

What you've just learnt applies to nozzles in Space.

Down in the atmosphere though, the story is slightly different because the atmospheric pressure (P_a) outside the rocket engine causes a pressure effect as shown here by the lighter-coloured double-shafted arrows:

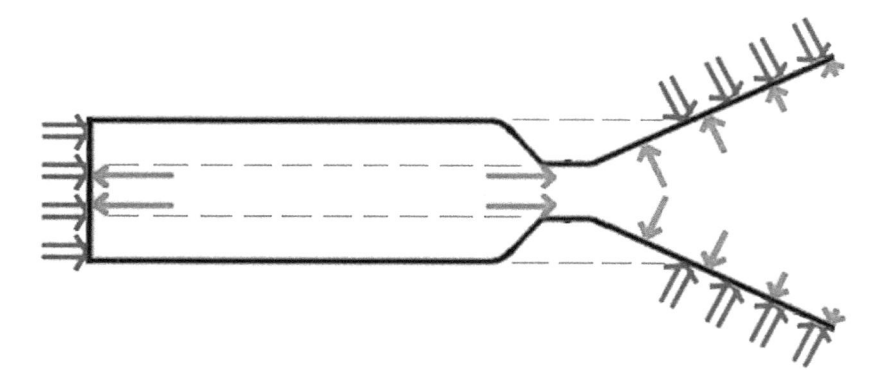

I'm only showing the unbalanced atmospheric pressure force (not cancelled-out by an equal force in the opposite direction). Note how it acts in the opposite direction to the (darker single-shafted arrows) thrust pressure so is called a **back-pressure**, and its equivalent horizontal area that it acts on is again the nozzle exit area including the throat. This back-pressure reduces the thrust so is detrimental.

The pressure thrust ($T_{pressure}$) is now equal to:

$$T_{pressure} = P_e A_e - P_a A_e = (P_e - P_a)A_e \qquad \text{equ. 5. 13}$$

so:

$$T = \dot{m}_{nozzle} C_e = \dot{m}_{nozzle} \left(V_e + \frac{(P_e - P_a)A_e}{\dot{m}_{nozzle}} \right) \qquad \text{equ. 5. 14}$$

Remember that the pressure of the exhaust gas drops steadily as the gas flows from the combustion chamber downstream through the nozzle:

Once the exhaust gas has expanded down to the same pressure as the atmosphere outside, there's nothing to be gained by adding extra nozzle. It won't do anything useful and is just dead weight.

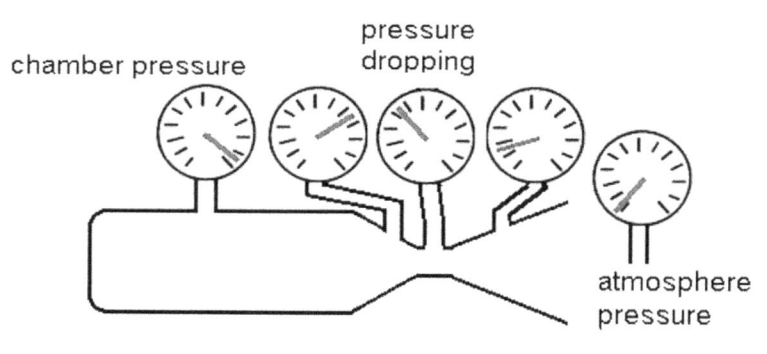

In fact it's been found by analysis that you get the highest thrust when the pressure thrust is zero, which, from the above equation, occurs when the nozzle exit pressure and the atmospheric back-pressure *are exactly equal*. With zero pressure thrust then the effective exhaust velocity is actually equal to the real exhaust velocity.

How far down the widening nozzle the gas reaches before it is equal to the pressure outside depends on the atmosphere's back-pressure (which gets larger the further down in the atmosphere you are), and the pressure in the combustion chamber.

So the exhaust flow's final exit Mach number depends on the difference between the pressure in the chamber and the pressure at the exit of the nozzle.

The bigger this difference ($\Delta P = P_{chamber} - P_e$), the higher the exhaust velocity.

The boundary layer

Now remember that I said on page 53 that pressure can't travel upstream against a supersonic gas flow. So whatever the back-pressure of the atmosphere is, this can't travel upstream to affect the nozzle wall's pressure distribution.

Unfortunately it turns out that the gas flow at the wall, sliding along the nozzle wall, gets slowed right down by the wall, simply because at the microscopic level, the supposedly smooth nozzle wall is very rough, and it effectively snags the flow sliding along it.

So there's a *very* thin layer of *subsonic* gas flow right at the wall, underneath the main supersonic flow.

This thin layer (called the **boundary layer**) is subsonic, so definitely *can* be upset by way too much back-pressure downstream. This is because the back-pressure can travel upstream along this boundary layer. This causes the flow to separate from the wall if the back-pressure is way too high, which means that the effective area of the nozzle is only the area of nozzle reached just before the flow separates.

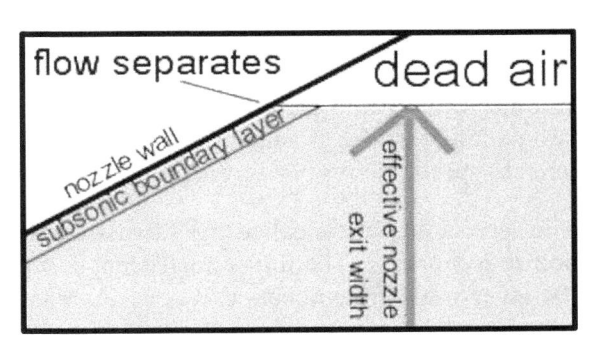

The separation point occurs when the gas pressure reaches a value of around:

$$P_e = \frac{P_a}{2.75} \quad \text{equ. 5.15}$$

Notice that this is lower than *Pa*, so you can get away with a *Pe* a bit lower than *Pa*; this could happen at launch (on the ground) when *Pa* is at its highest.

If *Pe* is lower than *Pa* then the exhaust is **over-expanded** (expanded too much). Conversely if *Pe* is higher than *Pa* then the exhaust is **under-expanded** because it could be expanded more by using a higher area ratio.

Exhaust velocity performance

You can calculate your rocket's effective exhaust velocity by rearranging the thrust equation as was done on page 47:

$$\textit{Effective exhaust velocity} = \frac{thrust}{nozzle\ mass\ flow\ rate} \quad \text{or:} \quad C_e = \frac{T}{\dot{m}_{nozzle}} \quad \text{equ. 5.16}$$

Once you've got this answer, you then consult rocketry textbooks or the internet to find out what value of effective exhaust velocity you *should* be getting from your rocket. You'll need to know the pressures at either end of the nozzle, and what type of propellants you're using.

If the exhaust velocity you're getting is a lot lower than it ought to be, then something's wrong: you're not getting enough thrust out of the pressure distribution in your nozzle. Probably you need to alter your nozzle, or maybe the mixture is wrong.

Actually, books and the internet might not give you the exhaust velocity straight away, they might give you values of a measure called **specific impulse** (I_{SP}). This is measured, rather oddly, in seconds, so that both Brits and Americans can agree on the values. (We usually use different sets of units, because the Americans haven't yet joined the civilised world and gone metric, but we both use seconds.)

The specific impulse includes the effect of the pressure thrust, so it gives you the effective exhaust velocity.

You get this simply by multiplying the I_{SP} (in seconds) by 9.81 in Britain (or 32.17 in America) which is numerically equal to one gee:

$$\textit{effective exhaust velocity} = I_{sp} \times 9.81 \quad \text{equ. 5.17}$$

If you find an *Isp* that's measured in metres per second rather than seconds then it's not really an *Isp*, it's actually the effective exhaust velocity because it's an *Isp* that's already been multiplied by 9.81

When you multiply *Isp* by 9.81 to get effective exhaust velocity, what you're also getting is a measure of how much thrust you'll get for every kilo per second of nozzle mass flow rate produced by the propellants. It's kind of like the 'miles to the gallon' or 'kilometres to the litre' measure that applies to cars.

Alternatively, because *Isp* is measured in seconds, then if the *Isp* was 400, that would mean that if you burnt 9.81 kilos of propellants in such a way as to get a constant one Newton of thrust from them, then they would keep producing this one Newton of thrust for 400 seconds.

Another way to find the effective exhaust velocity that rocketry textbooks or the internet might give you, is to use two parameters: one is called the **Characteristic velocity** C^* which is called a velocity *only* because it has the same units as speed, metres per second. Confusingly, it's got a star '*' even though it *isn't* the speed at the throat.

The second number is called the **Thrust coefficient or Force coefficient** C_F. The higher the number, the better the nozzle is working. The thrust coefficient describes the increase of thrust due to the pressure distribution pushing on the inner wall of the nozzle exit cone. Again, the thrust coefficient includes the pressure thrust so you get effective exhaust velocity.

Simply multiply them together:

$$\textit{effective exhaust velocity} = \textit{Characteristic velocity} \times \textit{Thrust coefficient} \quad \text{equ. 5. 18}$$

or:

$$C_e = C^* C_F \quad \text{equ. 5. 19}$$

Therefore the thrust is:

$$T = \dot{m}_{nozzle} C_e = \dot{m}_{nozzle} C^* C_F \quad \text{equ. 5. 20}$$

The sensible reason for rocketeers having two parameters to multiply together: $C^* C_F$ is that they have collected together all the factors that are affected by the combustion chamber only and *not* the nozzle and called it C^*. This lets you compare different rockets to see which is better without having to worry about the rockets having different nozzles.

Similarly, C_F is a collection of factors *only* affected by the nozzle and *not* the combustion chamber, so it lets you compare the effects of different nozzles.

They are defined by the throat area and chamber pressure thusly:

$$C^* = \frac{A^* P_{chamber}}{\dot{m}_{nozzle}} \quad \text{equ. 5. 21}$$

and:

$$C_F = \frac{thrust}{A^* P_{chamber}} \quad \text{equ. 5. 22}$$

so:

$$C^* C_F = \left(\frac{A^* P_{chamber}}{\dot{m}_{nozzle}}\right)\left(\frac{thrust}{A^* P_{chamber}}\right) = \frac{thrust}{\dot{m}_{nozzle}} = C_e \quad \text{equ. 5. 23}$$

Of more use is how they're derived (see appendix 3 for the derivations); giving us the following expressions:

$$C^* = \sqrt{\frac{R\,T_{chamber}}{\Gamma}} \qquad \text{equ. 5.24}$$

where R is the specific gas constant of the chamber gas.

$$C_F = \sqrt{\Gamma\left(\frac{2\gamma}{\gamma-1}\right)\left(1 - \left(\frac{P_{exit}}{P_{chamber}}\right)^{\frac{\gamma-1}{\gamma}}\right)} + \left(\frac{A_{exit}}{P_{chamber}\,A^*}\right)(P_{exit} - P_a) \qquad \text{equ. 5.25}$$

where the right-hand term deals with the pressure thrust as before.

Note that C^* only depends upon combustion chamber conditions alone, and C_F depends only on nozzle conditions.

In both the C^* and C_F equations:

$$\Gamma = \gamma\left(\frac{2}{\gamma+1}\right)^{\left(\frac{\gamma+1}{\gamma-1}\right)} \qquad \text{equ. 5.26} \quad \text{and is a constant.}$$

[Note that some books use a different equation for Γ but they're putting it into slightly different equations for C^* and C_F. The final answers work out the same though.]

C^* and C_F are also useful because you can use either one of them to quickly work out what nozzle throat area A^* you need for your rocket:

$$A^* = \frac{\dot{m}_{nozzle}\,C^*}{P_{chamber}} \qquad \text{equ. 5.27} \text{ (from equ. 5.21)}$$

or:

$$A^* = \frac{thrust}{P_{chamber}\,C_F} \qquad \text{equ. 5.28} \text{ (from equ. 5.22)}$$

Heating the gas

You might be thinking at this point that I've missed something out. The lengthy explanations I gave you in the previous chapter about nozzle pressure distribution, nozzle mass flow rate, and exhaust velocity, didn't care whether the exhaust gas was hot or cold. But most rockets go to all the bother of burning the propellants to heat up the exhaust gas, a huge jet of flame comes out the end of the nozzle.

What's the point of doing that?

You might think it's done to give us more thrust from the rocket, but actually, (assuming the exhaust gas is Ideal, and you don't change the combustion chamber pressure nor the throat area), the thrust doesn't increase if you heat the gas, it stays the same! So what is the point of hot gas?

What in fact happens is that the specific impulse (the measure of performance) of the engine increases. So less propellant is required to produce the same thrust.

Compare two gasses

From now on I'm going to compare two gasses. One is the gas in a cold gas thruster - a simple gas-powered rocket that doesn't heat the gas - and the other is the same type of gas, but it's been heated to the hot temperature of gas in a combustion chamber. The gas in the tank of the 'cold' gas thruster is at the Kelvin temperature of the air on a warm day so I'll refer to it as 'warm' gas. The hot gas I'll call 'very hot'.

We'll make both gasses flow down nozzles with the same pressure drop (chamber pressure minus nozzle exit pressure), and the same area ratio. This means that the Mach number of the gasses as they leave the nozzle exits will be the same for both gasses (see equation in appendix 3). The temperature makes no difference to this Mach number! But as we'll see shortly, just because the Mach numbers are the same doesn't mean the exhaust velocities have to be the same.

What heating *doesn't* do

First, let's investigate what heating the combustion chamber gas *doesn't* change:

First off, we'll assume that the gas is Ideal, so we'll ignore the small changes to the gas's chemistry as it's heated. All this would otherwise do is slightly alter the isentropic exponent γ which also alters the specific gas constant R of the gas (see chapter 1).

We can assume for now that once it's fixed within the combustion chamber, the gas's γ doesn't change as the gas flows through the nozzle. And we'll assume that both gasses have the same γ. A fixed γ means that for the same nozzle area ratio, the nozzle thrust coefficient C_F (see page 66 and appendix 3) is exactly the same whether the gas is warm or very hot.

The nozzle thrust coefficient C_F is the same warm or very hot because the pressure distribution along the nozzle wall *is exactly the same* warm or very hot as it depends only on Mach number (see the isentropic flow equations in appendix 3). And as I've said, the Mach number changes down the nozzle are exactly the same whether the gas is warm or very hot.

What heating *does* do

So if the thrust coefficient is the same, then the characteristic velocity C^* must be different to get the higher specific impulse.

Remember that C^* describes the combustion chamber alone, therefore whatever's causing the extra performance originates inside the combustion chamber. This is what we expect, as we're changing the gas in the combustion chamber by heating it.

So what could be making the extra performance?

Remember that I said earlier on page 51 that the nozzle mass flow rate gets smaller as the gas flowing through the throat gets hotter.

If the mass flow rate gets smaller, then surely the thrust gets smaller, because remember that thrust is equal to mass flow rate times exhaust velocity.

But hang on, I also said that the thrust remains the same whether the gas is hot or cold. Well, that can only occur if *the exhaust velocity gets larger by exactly the same factor as the mass flow rate gets smaller when you heat the gas*, so that when you multiply them together, there's no change.

That's exactly what happens: for example, if the mass flow rate halves, the exhaust velocity doubles. They both change as the square root of the Kelvin temperature of the gas trying to flow through the throat.

Earlier (equation 4.13) I showed that:

$$A^* = \sqrt{T^*} \left(\frac{\dot{m}_{nozzle}\sqrt{R}}{P^*\sqrt{\gamma}} \right) \quad \text{equ. 5.29}$$

Rearranging:

$$\dot{m}_{nozzle} = \frac{1}{\sqrt{T^*}} \left(\frac{A^*P^*\sqrt{\gamma}}{\sqrt{R}} \right) \quad \text{equ. 5.30}$$

where everything in the brackets is constant with temperature.

As we find from appendix 3, derived from enthalpy:

$$V_e = \sqrt{\left(\frac{2R\gamma}{\gamma-1}\right) T_{chamber} \left[1 - \left(\frac{P_{exit}}{P_{chamber}}\right)^{\frac{\gamma-1}{\gamma}}\right]} \quad \text{equ. 5.31}$$

Now:

$$T_{chamber} = T^* \left[1 + \left(\frac{\gamma-1}{2}\right) M^2\right] \quad \text{equ. 5.32}$$

from stagnation conditions (appendix 3), and at the throat the Mach number $M = 1$, so:

$$T_{chamber} = T^* \left[1 + \left(\frac{\gamma-1}{2}\right) 1^2\right] = T^* \left(\frac{\gamma+1}{2}\right) \quad \text{equ. 5.33}$$

so:

$$V_e = \sqrt{\left(\frac{2R\gamma}{\gamma-1}\right) T^* \left(\frac{\gamma+1}{2}\right) \left[1 - \left(\frac{P_{exit}}{P_{chamber}}\right)^{\frac{\gamma-1}{\gamma}}\right]} = \sqrt{T^*} \sqrt{R\gamma \left(\frac{\gamma+1}{\gamma-1}\right) \left[1 - \left(\frac{P_{exit}}{P_{chamber}}\right)^{\frac{\gamma-1}{\gamma}}\right]} \quad \text{equ. 5.34}$$

(again, everything within the larger square root sign is constant with temperature)

So nozzle mass flow rate is proportional to 1 divided by the square root of throat temperature (known as *inversely* proportional) which I've highlighted in equation 5.30, whereas exhaust velocity is *directly* proportional to the square root of throat temperature as highlighted in equation 5.34. As one goes up, the other goes down in exact proportion as the gas is heated. Multiply them together to get thrust, and the square root of throat temperature cancels out.

Now the temperature of the gas at the throat depends directly on the temperature of the gas in the combustion chamber:

From above:

$$T^* = T_{chamber} \left(\frac{2}{\gamma+1}\right) \quad \text{equ. 5.35}$$

It's the chamber temperature that's driving the whole thing, so it's the chamber temperature that's important.

So, the exhaust velocity increases as you increase the square root of the combustion chamber Kelvin temperature.

Okay, and so what?

Why we heat the gas

But consider this: by heating the gas, we've achieved the same thrust but with a lower nozzle mass flow rate. And remember that the nozzle mass flow rate is equal to the mass flow rate of propellants into the combustion chamber. So we've achieved the same thrust but by burning *a lot less propellant*.

And *that's* why we heat the gas!

It's fascinating: the gas in the combustion chamber expands (thins out) due to the heating; the collisions between the gas molecules are harder, so the molecules are bouncing further apart. There are now fewer molecules within every litre of chamber gas so the gas's density has lowered.

[As a useful side-effect, there are fewer gas molecules needed inside the chamber to get the pressure up, so the gas in the chamber has less mass which means the rocket has less mass (less inertia) so is easier to accelerate.]

So the gas density is now less within the very hot rocket. There are fewer molecules bouncing off the walls of the combustion chamber and nozzle, and yet the chamber pressure - and so the thrust - is still the same. This can only happen if each of the fewer molecules is hitting the wall harder, and this is of course what 'temperature' is, it's the average speed of the molecules moving within the gas. The molecules are hitting the wall faster, giving more of a momentum exchange per collision (momentum is the molecule's mass times its velocity of impact).

So the reason that one factor in the thrust equation ($\dot{m}_{nozzle}$) goes down with temperature by exactly the same amount that the other (Ve) goes up with temperature is because they're both intimately linked to the gas density; the density unites them both even though they are traditionally treated as separate entities.

But don't be fooled into thinking that because the density is lower that somehow because the very hot gas is very hot it expands more inside the nozzle than the warm gas, and that speeds up the very hot flow more than the warm flow in the nozzle because of mass continuity.

No, the gas equations for rocket nozzles are quite clear (see the stagnation temperature equation and the isentropic flow equations in appendix 3): the amount of expansion (the ratio of the density at the beginning of the nozzle over the density at the nozzle exit) depends only on the difference in Mach number between the beginning of the nozzle and the nozzle exit, and that Mach number difference hasn't changed one bit just because the flow is very hot.

Any expansion will of course increase the exhaust velocity, but the ratio of expansion happening within the nozzle is exactly the same whether the flow is warm or very hot.

However, just because the *ratio* of the density changes along the nozzle is the same doesn't mean that the actual density *values* are the same: 3/6 equals 2/4 but the numbers that form the fractions are different.

Combustion chamber temperature

The difference isn't what expansion happens in the nozzle, it's the expansion of the gas in the combustion chamber *before* the gas reaches the nozzle.

When this very hot low density gas then flows into the nozzle, it *has* to move faster at all points down the nozzle compared to if it were cooler and denser.

Why?

The (rearranged) mass continuity equation 1.18 can help us here:

$$v = \frac{\dot{m}_{nozzle}}{\rho A} \quad \text{equ. 5.36}$$

If the density ρ plummets (which it does) on the denominator of this fraction, then this greatly increases velocity v.

So the lower density of the very hot flow, which was achieved in the combustion chamber, has magnified the speed of the flow all along the nozzle and out the end of the nozzle: that's the reason that the exhaust velocity is faster.

Dropping temperature

We will investigate this increase in nozzle gas velocity further by investigating the definition of the speed of sound. However, firstly there's one new bit of information that you need to know. Because the enthalpy drops along the nozzle as it converts to kinetic energy, then the temperature plummets as gas flows down the nozzle (remember enthalpy is proportional to Kelvin temperature).

How much the temperature drops depends on what the Mach number of the flow is: the higher the Mach number gets inside the nozzle the more the temperature falls (see the stagnation temperature equation in appendix 3).

Both of the gasses we're looking at have the same Mach number at their throats (Mach 1) and also have the same higher Mach number at their nozzle exits, so the temperature of both gasses will drop by the same fraction. If the very hot gas gets 2/3rds cooler by the time it leaves the nozzle then the warm gas will also get 2/3rds cooler.

If the very hot gas chamber temperature was 3000 Kelvin then a 2/3rds drop will lower it down to 1000 Kelvin; a drop of 2000 K which is a large drop!

Let's say for easy counting that from now on we'll set the very hot gas combustion chamber temperature to be four times as hot as the warm gas chamber temperature.

Now the Mach number of the flow at any point you care to look at down the nozzles is the same for both nozzles. This means that the very hot gas will *always* be four times as hot as the warm gas at all points down the nozzle, and out the end of the nozzle, even though the temperatures are dropping. It'll always be four times as hot because it started four times as hot in the combustion chamber. (See the stagnation temperature equation in appendix 3).

So, starting in the combustion chamber and going downstream:

The very hot gas combustion chamber temperature is four times as hot as the warm gas chamber temperature. And so at the throat, the very hot gas is four times as hot as the warm gas because it was so to begin with in the combustion chamber. It's the chamber temperature that's driving the whole thing, so it's the chamber temperature that's important.

The speed of sound

So the two gasses have the same nozzle exit Mach number, but the very hot gas has four times the temperature of the warm gas as it leaves the nozzle.

Now the speed of sound (the value of Mach 1) changes with temperature. The hotter a gas is (in Kelvin), the higher is its speed of sound. The speed of sound is still Mach 1, but now Mach 1 is much faster.

So if the exhaust velocity of both gasses at their nozzle exits is Mach 3, this is three times a much higher speed of sound for the very hot gas, so the exhaust velocity is faster for the very hot gas.

How much faster?

The speed of sound a depends on the square root of the temperature:

$$a = \sqrt{\gamma R T} \quad \text{equ. 5.37}$$

where R is the specific gas constant of the gas, and T is the gas temperature in Kelvin.

As the very hot gas is four times the temperature of the warm gas at the nozzle exit, then its speed of sound at the nozzle exit is the square root of 4 = 2 times higher, so the exhaust velocity of the very hot gas is 2 times faster than the warm gas.

The nozzle exit temperature is four times hotter because the combustion chamber temperature is four times hotter.

So if the combustion chamber temperature is four times hotter, the exhaust velocity is $\sqrt{4}$ = 2 times faster.

Similarly, if the chamber temperature is 3 times hotter, the exhaust velocity is $\sqrt{3}$ times faster.

So we've discovered an exhaust velocity rule: the exhaust velocity increases as the square root of the combustion chamber temperature increases.

So the difference in exhaust velocity between the warm and very hot rocket depends upon the square root of the combustion chamber temperature.

Sure enough, the equation for the characteristic velocity C^* we saw earlier depends upon this too: C^* increases with the square root of the combustion chamber temperature:

$$C^* = \sqrt{\frac{R\, T_{chamber}}{\Gamma}} = \sqrt{T_{chamber}} \sqrt{\frac{R}{\Gamma}} \quad \text{equ. 5.38}$$

(where R is the specific gas constant of the chamber gas)

We just picked the number 4 for easy counting. Actually, for a Nitrous oxide hybrid rocket engine, the gas is around 12 times hotter (in Kelvin) than room-temperature (warm) gas. So the thrust is $\sqrt{12}$ = 3.46 times more than a warm gas. That's a big improvement.

We'll talk more about this shortly.

Liquid propellant rockets burn very hot and therefore give the highest exhaust velocity of the different types of rockets I've mentioned, followed by hybrids, and then solids.

Cold gas thrusters (warm gas) give a much lower exhaust velocity than for liquid rockets, but they're still used as they're very simple to build and operate.

A thrust loss

But hang on, if the temperature of both gasses drop down the nozzle then surely this will increase the density of both gasses?

Sadly this is true, the lower temperatures will increase the gas density at the nozzle exit because the gas molecules are moving more slowly and therefore aren't bouncing so far apart, so the expansion (reducing density) downstream of the nozzle throat will be badly affected: the exhaust velocity will end up lower for both gasses: they'll be affected equally badly.

To find out how that happens, we can consult the Ideal gas equation 1.14.

Let's pick some simple numbers to show this equation doing just that. Make $P = 8$, $R = 1$, and $T = 1$ for the warm gas. So:

$$\rho = \frac{P}{RT} = \frac{8}{1 \times 1} = 8 \quad \text{equ. 5. 39} \quad \text{so the density of the warm gas is 8.}$$

Now the very hot gas is four times as hot as the warm gas, so give it a temperature of 4. So:

$$\rho = \frac{P}{RT} = \frac{8}{1 \times 4} = 2 \quad \text{equ. 5. 40} \quad \text{so the density of the very hot gas is indeed lower.}$$

The Ideal gas equation is useful because it tells us is how much lower. The density of this very hot gas is a quarter of that of the warm gas.

Now let's use the Ideal gas equation to investigate the nozzle downstream of the throat. We know that the gas expands (the density drops) because the pressure drops. How will the drop in temperature down the nozzle interfere with this?

Pick the same easy numbers again: make $P = 8$, $R = 1$, and $T = 1$ for the gas. So:

$$\rho = \frac{P}{RT} = \frac{8}{1 \times 1} = 8 \quad \text{equ. 5. 41} \quad \text{so the density of the warm gas is 8 again.}$$

Now to account for the temperature dropping down the nozzle, divide the temperature by 4, so $T = ¼$, but reduce the pressure even more, divide it by 8 so $P = 1$. So:

$$\rho = \frac{P}{RT} = \frac{1}{1 \times \frac{1}{4}} = 4 \quad \text{equ. 5. 42}$$

So the density of the gas has still dropped, it's still expanded, which will increase the exhaust velocity. What we did to get this to happen was to drop the pressure more than the temperature dropped. The flow down a real nozzle does exactly that.

If there was some way to keep the temperature the same all the way down the nozzle then we'd get lots more expansion (try the same sum but keep $T = 4$) but sadly there isn't because the enthalpy is getting used up. [Unless you can think of a cunning way of adding progressively more energy to the nozzle flow from some other energy source.]

Will the chamber pressure have to change?

All this maths is all very well, but will we need more chamber pressure to get the very hot gas up to higher speed?

Fortunately not, because a scientist called Daniel Bernoulli discovered that if a gas's density is lowered, then it's easier to accelerate it.

This means that the same chamber pressure can accelerate the very hot exhaust gas to a higher speed compared to when it was warm because it's now less dense.

The exhaust velocity is therefore faster than with a warm gas for the same chamber pressure.

Chapter 6: Energy and exhaust velocity

In the following chapters we'll design a rocket engine that could be used to take you in a spacecraft up above 100 kilometres (into Space). [The same principles apply to a small hobbyist/HPR rocket.]

We'll launch from a balloon floating at 21 kilometres up so that the atmosphere's back-pressure on the nozzle is low, therefore the nozzle's thrust (and the effective exhaust velocity) is higher.

Also, the spacecraft isn't slowed nearly so much by aerodynamic drag as it passes through the air, because the air is very thin above 21 kilometres.

We'll look at designing a hybrid rocket, as it's easier to make and test, and is safer to operate, than a solid or liquid rocket. If one day you go into Space either in your own spacecraft or as a tourist in someone else's, chances are the rocket will be a hybrid, and the chances are it'll use Nitrous oxide as its oxidiser as Nitrous is easy to work with.

A Nitrous hybrid is a beautifully simple type of rocket; it's easy to knock one up and it'll usually do what it's supposed to (though there are some safety issues that I'll talk about in chapter 8). But to get the most out of the hybrid you have to tune it up carefully otherwise you could lose half the thrust.

Nitrous oxide hybrids are simple to make, but there's a lot going on when they're burning.

The first thing to look at *is* the burning: we need to know how much *energy* we can get out of the propellants.

So what are the energies that occur inside the combustion chamber?

Chemical energy

When the propellants were created, chemical potential energy was stored inside the molecules of the propellants in the same sort of way that electrical energy is stored in a battery. When we burn the propellants we can get this energy out as heat energy which heats the gas in the combustion chamber.

That's the simple explanation, and you might just want to leave it there and skip some of the following paragraphs, but that's the trouble with talking about energy, it doesn't explain anything. Most science books just leave it there, and go off on a sidetrack into chemistry.

I've had the energies explained to me in a bit more detail, though bear in mind that when we describe the inner workings of atoms we can *only* discuss them in terms of energy simply because nobody really knows exactly what's going on. That's because it's physically impossible to build a microscope powerful enough to actually see atoms, and even if we could, the individual particles of the light we would use to see the atoms are atom-sized so would collide with the atoms and upset our experiment. [Electron microscopes can just detect the fuzzy outlines of atoms, but they're only seeing the clouds of electrons around the atom.]

So science is literally groping in the dark when it tries to explain atoms. The fact that we can understand anything at all about them is therefore remarkable.

When the propellants were created, the propellant atoms formed into many clumps of two or three atoms called molecules, and they're 'glued' together by a **chemical bond**. These chemical bonds are strong attractive forces made of static electricity in the same way that when a plastic comb is charged electrically by combing your hair, it can make tiny bits of paper stick to it. The electricity is creating a magnet-like attracting force called **electrostatic force**.

The atom

A simple picture of an atom is a central lump called the **nucleus** which is positively electrically charged. Several negatively electrically charged electrons orbit around the nucleus in several separate orbits like planets orbiting the Sun at different distances (orbital 'heights') from the nucleus.

It so happens that the negative charge from all the electrons exactly equals and cancels out the nucleus's positive charge, so that the atom as a whole has zero overall electrical charge. But if the atom gains or loses electrons then its zero charge will change to either positive (it has too few electrons) or negative (it has too many electrons) so will want to electrostatically stick to other atoms.

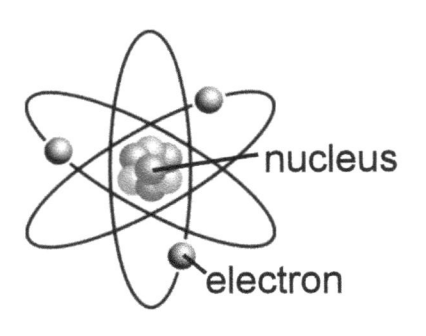

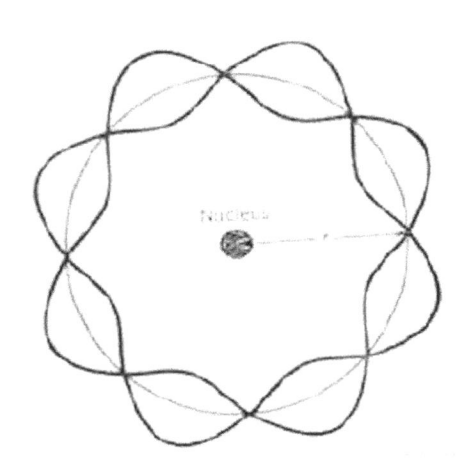

More recently, quantum theory modifies this picture: The old picture of electrons circling the atomic nucleus like little planets did not fit the new data physicists were gathering. The scientist Schrödinger proposed a new theory that replaced orbiting electrons with an image more like standing waves around the nucleus: wavy orbits like the waves found in a plucked guitar string.

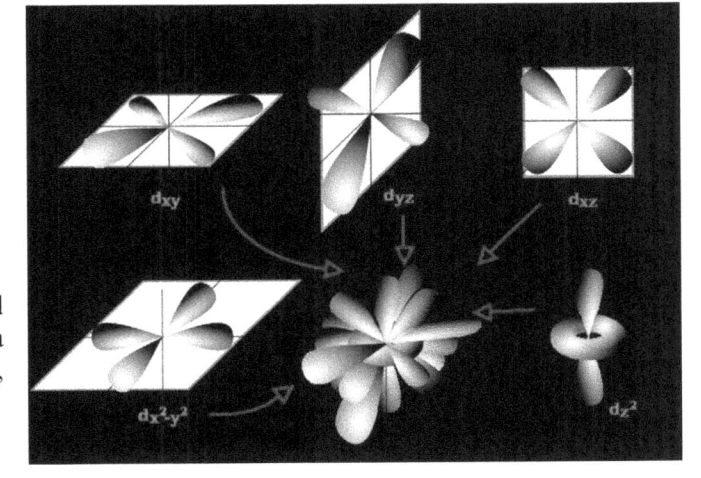

This evolved into the idea of 'orbitals'; symmetric lobed regions stretching out from the nucleus where there was a high probability (but not a certainty) of finding an electron, as shown here:

Chemical bonds

The important thing to remember about chemical bonds are that they're rather a fiction:

"Sometimes it seems to me that a bond between two atoms has become so real, so tangible, so friendly, that I can almost see it. Then I awake with a little shock, for a chemical bond is not a real thing. It does not exist. No one has ever seen one. No one ever can. It is a figment of our own imagination." So said C.A. Coulson, who played a central role in the development of quantum theories of chemical bonding.

It is more useful to regard a chemical bond as an *effect* that causes certain atoms to join together to form enduring structures that have unique physical and chemical properties.

There are several types of chemical bond, but they all use electrons to 'stick' the atoms together, and it's only the electrons furthest from the nucleus, known as **valence electrons**, that are involved.

For example, if certain atoms are brought close together, then one valence electron from one atom might have an orbital that overlaps with the orbital of a valence electron from its neighbouring atom. For rather obscure mathematical reasons, this can cause the two negatively charged electrons to spend a lot of their time positioned between the two nuclei (which are both positively charged).

In practice, what is happening is that as two atoms approach each other, the chances increase that a negatively charged electron from one will find itself attracted to the positively charged nucleus of the other, and the same thing would happen from the other side.

At a certain point, the two electrons begin jumping back and forth between the two nuclei, creating an electron exchange at a rate of billions of times per second. In a sense, the two electrons can't tell which nucleus they belong to. This exchange creates an 'exchange energy' that has to be overcome to break the bond.

Then both nuclei will be electrically attracted towards this 'bonding pair' of electrons, because opposite charges attract each other. This attraction causes a **covalent** chemical bond: the two nuclei are drawn towards each other. The greater the overlap of orbitals, the more exchange energy is created, and the stronger the bond.

Alternatively, an atom can simply capture another atom's valence electrons, making one atom positively charged and one negatively charged, so they attract each other. This is an **ionic** chemical bond.

Generally, a bond is neither completely covalent nor completely ionic (or the other weaker types) but is simply more one than the other.

Linus Pauling, who successfully applied quantum mechanics to describe bonds, proposed that when certain criteria were met, resonance could exist between the ionic and covalent forms of a molecule. For example, hydrogen chloride could be viewed either as a hydrogen atom linked to a chlorine atom through a purely covalent bond, or as a positively charged hydrogen ion and a negatively charged chloride ion linked with a purely ionic bond. The actual molecule, Pauling said, is a sort of hybrid; a structure that resonates (oscillates) between the two alternative extremes.

[An ion is an atom or a molecule in which the total number of electrons is not equal to the total number of protons in the nucleus, giving the atom or molecule a net positive or negative electrical charge.]

Since this electrical attraction in a chemical bond is so 'attractive,' why doesn't the resulting molecule collapse totally in on itself to become a little black hole? Fortunately there are repulsive forces at work that also become larger as the pieces (electrons or protons) get closer together. (Even modern quantum theory has to accept this.) So an equilibrium finally exists where attraction and repulsion balance at a certain distance called the **bond length**. (This length correlates with the strength of the bond). The full picture has the bond in oscillation about an equilibrium point, resulting in observable absorption and emission of electromagnetic radiation: more on that shortly.

So the chemical bonds are formed from electrical energy (electrostatic force) so in a way, all chemical rockets are electrically powered!

Combustion

When we burn the propellants inside the combustion chamber, the chemical bonds between the atoms making up the molecules break and rearrange. This lets the electrical energy out.

It's important to realise that the *breaking* of chemical bonds never releases energy. Energy is only released when chemical bonds are *formed*. In general, a chemical reaction involves two steps: 1) the original chemical bonds between the atoms are broken, and 2) new bonds are formed. These two steps are sometimes lumped into one event for simplicity, but they are really two separate events:

Bond-*breaking* is an **endothermic** process: energy has to be absorbed from the surroundings to break the bonds.

It's not quite as obvious that bond-*making* is an **exothermic** process (energy is given out to the surroundings), but if a particular bond in a molecule is broken and then reformed, the same amount of energy must be involved in both processes, because of the Law of conservation of energy.

This released electrical energy then changes into heat energy (see below) which heats the gas in the combustion chamber.

'Burning' (properly called combustion) is a chemical reaction with oxygen called **oxidation**. The oxidiser (which contains oxygen) **oxidises** the fuel. It changes the propellants (called the **reactants**) into new chemicals (called the **products**).

Oxygen is highly electronegative, which means that it really likes to steal electrons from the fuel molecules if given the chance, in order to form new bonds.

But first we need to supply some 'activation' energy to break the first bonds to get the combustion started, usually heat from an igniter. And oxygen is a bit odd in that it suddenly becomes much more chemically reactive when its temperature is raised because then its valence electrons start to behave differently: they then want to form bonds, whereas they didn't when cold.

Most rocket fuels, such as kerosine (paraffin), are made from large molecules that are long chains of bonded carbon atoms (C) fringed with bonded hydrogen atoms (H) so they're called **hydrocarbons**.

Each short line in this diagram represents a chemical bond:

```
        H              H H            H H H
        |              | |            | | |
    H — C — H      H — C — C — H  H — C — C — C — H
        |              | |            | | |
        H              H H            H H H
     Methane          Ethane          Propane
```

```
    H H H H H H H H H H H H H H H H
    | | | | | | | | | | | | | | | |
H — C — C — C — C — C — C — C — C — C — C — C — C — C — C — C — C — H
    | | | | | | | | | | | | | | | |
    H H H H H H H H H H H H H H H H
                    Kerosine
```

A chemical bond oscillates: you can think of a bond between two atoms as a simple spring with two heavy balls at either end. It vibrates at a fixed frequency that is unique to the chemistry of each bond (the type of atoms at either end of it). [The sense we call 'smell' picks out these unique frequency signatures to identify the substance.]

The heat from the igniter, or the continuing heat from the combustion, makes the hydrocarbon chains oscillate so violently that their 'springs' break; they splinter into smaller molecular fragments. This is termed **pyrolysis**.

Now that these fragments are free, the oxygen steals some of their electrons, and this causes new bonds to be made between the fragments and the oxygen atoms. These new bonds need much less electrical energy to stay together than the original bonds, so the excess energy leaks out eventually as heat.

But how does the chemical bond's electrical energy turn into heat?

This involves a branch of physics called quantum electrodynamics.

As far as I understand it, there's a force-field that pervades all of Space called the **electromagnetic field**. It interacts with electrical particles such as electrons. This causes their electrical attracting (electrostatic) force which, as we've seen, can be used to make the chemical bonds.

Energy is released when bonds are formed because the electrons involved move into a lower energy configuration in the bonded arrangement than they had before the bond formed (a lower 'orbital height' around the atoms if you like). As the electrons lose energy, they dump this energy into the electromagnetic field. This somehow (nobody knows how) causes the field to beam out pulses of **electromagnetic radiation** just as a radio transmitter antenna does, with an energy equal to the loss of the electron's energy.

We know electromagnetic radiation as light; brief pulses or **quanta** of light come out of the molecules. A 'quantum' (one quanta) is a tiny packet of energy; quanta of light are called **photons**.

The photons of light energy that radiate out from the molecules can be absorbed by other atoms/molecules, making them vibrate with more kinetic energy. In this way, heat is transferred by radiation.

If this heat transfer raises the (vibrational) energy of the molecule above a certain point, its chemical bonds ('springs') are broken, causing yet more light to come out. So that's why burning keeps going after ignition, it's spontaneous: as long as there are chemical bonds still available to be broken and reformed (as yet unburned propellant), then more and more light (and heat) comes out.

Sometimes we can see this light, (the yellow colour of a flame is caused by the light given out by carbon soot atoms), but in the main it's at a frequency that our eyes can't see (infrared radiation).

So the electrical bond energy transforms into light energy. Then some of the light energy stays deep inside the gas and makes the gas atoms shake with more kinetic energy.

Now this describes transfer of energy purely by radiation. However, when a slow-moving 'cold' molecule collides with one that is vibrating rapidly, it rebounds at a higher speed. As its speed is what we define temperature as, it is now a 'hot' molecule.

And then simple collision between this fast-moving 'hot' molecule and a slow-moving 'cold' one causes a momentum exchange that speeds the slow one up somewhat: this is called heat transfer by conduction. With billions of moving particles colliding into each other (molecules, bare atoms, and ejected 'free' electrons), an area of high particle kinetic energy will slowly transfer through the gas until thermal equilibrium is reached (the average speed, or temperature, is the same throughout the gas).

This 'heat' energy inside the gas is internal energy (chapter 1); an internal store inside the combustion chamber that can be used later as enthalpy energy to expand the gas down the nozzle.

The enthalpy of the propellants

So if burning released a kiloJoule of chemical energy, that turns into a kiloJoule of enthalpy, which is a potential energy that we can then use to do something useful. (Yes, heat and light also come out of the combustion chamber gas, but these energies are tiny compared to the enthalpy, so can be ignored.)

In the combustion chamber, the gas as a whole is stagnant, so its speed squared is very low, which makes its kinetic energy practically zero. Then within the nozzle, the enthalpy energy gets converted into kinetic energy, which is the energy of the exhaust velocity (squared) as the gas leaves the end of the nozzle.

Sadly we can't change all the enthalpy into kinetic energy. How much we can get out depends on the difference in pressure between the combustion chamber and the nozzle exit, the bigger this difference the more we can get out. So the exhaust gasses are still hot, perhaps 1000 Kelvin, as they leave the nozzle. They still contain some enthalpy that couldn't be changed into kinetic energy.

Perhaps you can think of a way of harnessing this wasted enthalpy. After all, you can stick a turbocharger into a car's exhaust to do something useful. Unfortunately, the exhaust gas from a rocket is very much hotter, so tends to melt anything you put into it.

Actually, using just a nozzle, we could never get all the enthalpy out even if we used a huge nozzle that expanded right down (in pressure) to nearly the vacuum of Space. As the gas cooled as its enthalpy dropped down this huge nozzle, it would eventually get so cold that it would condense into a liquid: a hail of droplets. A load of liquid droplets can't exert a pressure on the nozzle like a gas can.

Sometimes in real rockets a chemical in the exhaust gas does turn into a liquid; it slows the rest down so **condensed species** as they're called are a hindrance.

So that's energy; if we know how much energy came out of the propellants, we can work out what the exhaust velocity will be (from its kinetic energy) so that we can then calculate the thrust. It really is that simple, we don't have to bother to learn how it happens.

But there's one major problem:

Finding out exactly how much energy came out of the propellants during combustion turns out to be computationally difficult. That's because so many things are happening during the combustion process.

Heat energy is *used up* as the propellants burn and their molecules break up into separate atoms. Actually it's an enthalpy, called the **enthalpy of the reactants** because the burning is having to work against the force of the combustion chamber pressure.

But similarly, energy is coming out of the propellants as the atoms re-group into the new molecules that make up the exhaust gasses. Again this happens against the chamber pressure so this is also an enthalpy, called the **enthalpy of the products**.

The **heat of reaction** is the name given to the difference between the enthalpy of the reactants and the enthalpy of the products, so it should really be called the enthalpy of reaction [sometimes it is].

$$Heat\ of\ reaction = H_{reactants} - H_{products} \quad \text{equ. 6.1}$$

We choose propellants that have a big $H_{reactants}$ but a small $H_{products}$ as that'll give the biggest $H_{reaction}$.

I use the rocketeer's version of this equation:

$$H_{reactants} - H_{products} \quad \text{equ. 6.2}$$

but if you're a chemist, you might be used to this equation being the other way around:

$$H_{products} - H_{reactants} \quad \text{equ. 6.3}$$

If rocketeers use the chemist's version, they'll get an answer that is negative. The chemist would then say that the negative sign in front of your answer simply denotes that heat came out of the combustion (exothermic) rather than you having to put it in (endothermic) but of course we knew that.

So we first need to find both $H_{products}$ and $H_{reactants}$.

$H_{reactants}$ are easy, they're listed in books and on the internet, look up '**heat of formation**' or '**enthalpy of formation**'. (These are usually specific quantities.)

But $H_{products}$ depend on many things such as the mixture, and what exactly the chemicals of the products are. But what exactly these products are depends on how hot the gas gets, but how hot the gas gets depends on what these products are: it's a recursive situation.

Thermochemistry

I won't attempt to hand-calculate the combustion thermochemistry in a hybrid rocket as it's onerous as we shall shortly discover. Instead there are several freely available computer programs that can do it for you in microseconds; one of which I'll discuss shortly. However, you're doubtless curious about how it's worked out by the software.

Reference enthalpies

We previously defined specific enthalpy as:

$$h = C_p T \quad \text{equ. 6.4} \quad \text{(strictly for an Ideal gas, where } Cp \text{ is the specific heat capacity at constant pressure.)}$$

Cp is measured empirically, but it's impossible to get *Cp* data at low temperatures, so if there was some odd behaviour at low temperatures, we'd never know about it. So *Cp* data stops below a warmish temperature. But that's not a problem, because we don't concern ourselves with low-temperature combustion in rocketry.

And furthermore, we never need to know the *absolute* value of enthalpy in our calculations, only the *difference* in enthalpy between products and reactants.

So what is done is to arbitrarily define reactant enthalpies at some internationally-agreed standard conditions (1 Bar and 25 degrees C, earlier analyses were at 1 atmosphere and 15 degrees C) as our 'zero' in the calculations, and work the *changes* in enthalpy from that point. Provided we only deal in enthalpy *differences*, then the answers work out correctly.

The standard heat (enthalpy) of formation *Hf* is the enthalpy change required to form one mole of the substance from its elements at the above standard conditions. The most stable form of an element (be it solid, liquid, or gas) that occurs at these conditions is *defined by us* to have an *Hf* of zero.

So oxygen or hydrogen gas at 1 Bar and 25 degrees C are given an *Hf* of zero, whereas water vapour H_2O is built from other elements, so is non-zero at -241.93 kiloJoules per gram-mole. Values for a host of other elements and compounds can be found in standard rocketry textbooks.

Enthalpy of reaction calculations are now straightforward, as a simple example illustrates:

$$2 \text{ moles } H_2(gaseous) + 1 \text{ mole } O_2(gaseous) = 2 \text{ moles } H_2O(gaseous) + heat \quad \text{equ. 6.5}$$

$$\Delta H^0_{reaction} = 2 \text{ moles } H_2O(gaseous) - [2 \text{ moles } H_2(gaseous) + 1 \text{ mole } O_2(gaseous)]$$

$$= 2 \times (-241.93) - (2 \times 0 + 1 \times 0) = -483.86 \text{ kiloJoules} \quad \text{equ. 6.6}$$

In this case, the –ve sign denotes an exothermic reaction.

This example is straightforward, because we knew - from basic chemistry - what the product would be (water vapour). However, in a complex reaction, we need a way to determine what the numerous products will be, and in what state and proportions they will occur.

Chemical equilibrium

But it's worse than that!

In the above chemical formula we assumed - from basic chemistry - that the reaction went to completion: all the hydrogen and oxygen became water. However, in the real world, this isn't quite true: there is no 'complete' reaction inside a closed container, but we can often make such an assumption to reasonable accuracy.

But: <u>*any chemical system in a closed container will always reach a state of chemical equilibrium*</u>. Sometimes that equilibrium state may be such that the container has *almost* no reactants in it but is still in equilibrium. A rocket combustion chamber is *almost* a closed container because the narrow throat keeps the products within the chamber for long enough to set up reverse reactions.

What is chemical equilibrium? It's the state of a reaction mixture at which the forward reaction rate is equal to the reverse reaction rate, so reactants become products, but oddly, some products also become reactants at the same time!

What equilibrium state the system ends up in *doesn't* depend on the amount of reactants: some equilibrium systems end up having mostly reactants, others, mostly products, and in rare cases, the same amount of reactants and products.

Upon ignition, there are only reactants in the combustion chamber, but as time progresses, products are formed and reactants are used up. When chemical equilibrium is finally reached, the amounts of reactants and products no longer changes with time, but are not necessarily equal to each other.

What is chemical equilibrium trying to tell us? Do we not burn up the propellants? Yes we do, pyrolysis still occurs, and the broken bits get vaporized into gas. It's simply that the chemistry in the combustion chamber rapidly gets horribly complex as reactions go back and forth.

For example, all three gasses in the above example (hydrogen, oxygen and water) would be present in the combustion chamber in unchanging proportions that depend upon the pressure, temperature, and initial mixture ratio.

And we also get side reactions occurring: extra unexpected reactions that try to bring the system towards chemical equilibrium (which they succeed in doing).

For example, our simple combustion of hydrogen and oxygen would end up being:

1. $2H_2 + O_2 \rightleftarrows H_2O$ Main reaction
2. $O_2 + H_2 \rightleftarrows 2OH$ Side reaction (two hydroxyl radicals formed: these have unpaired valence electrons)
3. $H_2 \rightleftarrows 2H$ Side reaction (dissociation and recombination: splitting and rejoining of molecules)
4. $O_2 \rightleftarrows 2O$ Side reaction (dissociation and recombination)

The double arrows show that the reverse reaction is also occurring.

So chemical equilibrium occurring within the combustion chamber is onerous. On the other hand, we can define chemical equilibrium mathematically, which allows us to then work out exactly which reactions *will* occur, and in what mole proportions. So we can, in fact, calculate what the gas passing out of the combustion chamber will be composed of.

Gibbs free energy

In thermochemical software, the method used to home-in on chemical equilibrium within the combustion chamber is **the minimisation of Gibbs free energy** (or chemical potential).

At the beginning of the 20th Century, J.W. Gibbs derived another property of gasses by combining enthalpy (*H*), temperature (*T*), and entropy (*S*):

Gibbs free energy:

$$G = H - TS \quad \text{equ. 6.7}$$

For a mixture of *n* chemicals:

$$G = \Sigma(G\,n) \quad \text{equ. 6.8}$$

which means 'the total sum (Σ) of the Gibbs free energy for each of the *n* chemical species'.

Gibbs free energy can be thought of as the driving energy that will cause a chemical to enter into a chemical change.

Using data at standard conditions (1 Bar, 25 C), the standard-state Gibbs free energy of reaction is the sum of the Gibbs free energies of formation of the products minus the sum of the Gibbs free energies of formation of the reactants (remember that the '0' means 'at standard conditions'):

$$\Delta G^0 = \Sigma \Delta G^0_{f\_products} - \Sigma \Delta G^0_{f\_reactants} \quad \text{equ. 6.9}$$

Tables of Gibbs free energies of formation exist which can be referred to.

Now it so happens that at chemical equilibrium $\Delta G^0 = 0$, and a graph of ΔG^0 versus mole concentration n bottoms-out at a minimum. So the software would vary the mole proportions of all the main reactions and side reactions it expects to occur (it's given some intelligence to guess which will occur) until the free energy graph curve is minimised at $\Delta G^0 = 0$

Energy balance

The software is almost there, however, the proportions of the individual gasses derived from minimising Gibbs free energy change with temperature, so the temperature of the gas mix in the combustion chamber now has to be calculated.

Fortunately, this is reasonably straightforward using the conservation of energy principle.

The enthalpy of reaction calculations described earlier assume that the products are formed at the standard temperature of 25 degrees C. The heat released is now used to heat the gas mix up to an initial guess at the chamber temperature.

- First, calculate the heat of reaction, using the minimisation of Gibbs free energy to determine the mole concentrations of the constituent gasses in the mix.

- Then add the energy released as the reactant parts of the mix are raised from 25 degrees C to the assumed combustion temperature.

- Now subtract the heat absorbed by the products parts of the mix as they're raised from 25 degrees C to the assumed combustion temperature.

- Is the answer zero (an energy balance)? If not, alter the guessed temperature and try again: if the answer was +ve, increase the temperature. If –ve, decrease the temperature.

Bear in mind that the definition of 'reactions' as compared to 'products' is fuzzy, as many reactions can go in the reverse direction.

PROPEP

Fortunately, NASA and similar organisations have written software to perform these calculations for you, and they've kindly put them on the internet.

One such program is called **PROPEP 3** (The PROPellant Evaluation Program version 3) and that's the one I use. The original PROPEP is showing its age; its descendant CEA is generally used now. However, it's much easier to add new propellants to the list contained within PROPEP.

PROPEP 3 isn't terribly user-friendly because it generates a 'results' file that is hard to understand, and whose answers use a lamentable mixture of metric and American units because the Americans haven't yet joined the civilised world and gone metric. (PROPEP measures $H_{reactants}$ in calories per gram instead of Joules per kilogram which is a nasty clash of units.)

Despite all that it's very useful. It does more than just work out the heat of reaction, and the minimisation of Gibbs free energy, it then simulates changing this heat into the gas's enthalpy within the combustion chamber, and then simulates changing this into the kinetic energy of the exhaust.

So it works out the exhaust velocity, as well as giving lots of other useful information about the gasses in the combustion chamber and also in the nozzle throat and nozzle exit.

When it was written in the 1980's, each run of PROPEP could take a few hours. With modern computers such as your one at home, it now takes fractions of a second.

You can download PROPEP 3 from Dave Cooper's site www.tclogger.com (click on the 'PROPEP 3 now available' icon).

I've put explanations of what the results file mean into a file that you can download from my rocketry society's website www.aspirespace.org.uk/technical_papers.html Consult appendix 4 for details.

I'll give you some PROPEP 3 results later in this chapter and also in appendix 2.

Propellant selection

In order to begin the PROPEP calculations we first have to choose a fuel, something that burns exothermically.

In chapter 2, High Density PolyEthylene burning in Nitrous oxide was selected as a best choice of propellant reactants.

We'll burn the HDPE and the Nitrous at the best mixture that gives the highest exhaust velocity, which the PROPEP 3 software says is a little over 6.4 kilograms of Nitrous for every 1 kilogram of plastic fuel. You need that much Nitrous because it's only 1/3rd oxygen in composition.

Actually, there are two ways to work out the energy within the nozzle. One way is to assume that whatever the mix of exhaust gas chemicals there are going into the nozzle, then this mix stays the same all the way along the nozzle. This is called **frozen flow** because the mix is 'frozen', that is, unchanging.

The other way is to assume instead that the mix is constantly changing; new molecules form and others break up as the temperature drops down the nozzle. This is called **shifting flow** because the mix shifts, (also known as **equilibrium flow**).

The best mixture of propellants (a ratio of 6.4 to 1) I mentioned above is the one that gives the highest exhaust velocity assuming frozen flow. For shifting flow it's 8.1 to 1.

Although PROPEP gives both frozen flow and shifting flow results, most rocketeers use the frozen flow results as they're easier to work with (because the mix doesn't change down the nozzle).

We'll be using frozen flow results from now on. Also, the shifting flow results are a bit optimistic, they give you a higher exhaust velocity than you'll ever actually get.

The actual best mixture is about halfway between the frozen and shifting mixture ratios, so we'll use a mixture of just over 7 to 1 in our hybrid engine design, but we'll run PROPEP at the frozen result of 6.4 to 1 which gives, for every one kilogram of propellants burnt: 865 grams of Nitrous to 135 grams of HDPE.

Liquid or gas?

Now before we start using the PROPEP 3 software, we've got to alter it.

PROPEP uses an input file that contains an extensive list telling it the enthalpy of formation of the reactants $H_{reactants}$ (see page 79) for many different propellants.

Included in this list is the enthalpy for Nitrous oxide, but, although it doesn't say so, it's actually the enthalpy for Nitrous oxide *gas* (vapour).

Now, when the Nitrous flows through any decent injector on its way into the combustion chamber, the majority of the Nitrous liquid flashes into vapour within the injector because of the sudden pressure drop within the injector (more on that later). However, when it enters the combustion chamber, it still has the *enthalpy* it had when it was still a liquid.

So we need to know the enthalpy of Nitrous liquid. We get this by calculating how much energy would be required to change Nitrous liquid into Nitrous gas whilst keeping the temperature constant, called a phase change.

This energy requirement is called the **specific enthalpy of vaporization** (Or specific heat of vaporization, see chapter 1) and for Nitrous oxide (at the same 25 degrees C) it has a value of 146.5 kiloJoules per kilogram. We now subtract this specific enthalpy of vaporization from the (specific) enthalpy of formation of Nitrous gas, to reverse the vaporization to get liquid from gas. This makes the heat of reaction lower, and so the effective exhaust velocity is lower.

Appendix 4 tells you how to alter PROPEP to input this *liquid* Nitrous, but be careful or you'll break the PROPEP software. (I've already done this alteration for you in a file you can download from my Aspirespace website; see appendix 4)

[As the Nitrous vaporizes within the injector, the energy required for the vaporization - the specific enthalpy of vaporization - is robbed from the Nitrous itself, and so the Nitrous entering the combustion chamber is considerably colder: it has a lower internal energy. Instead of using PROPEP results for liquid Nitrous at 25 degrees C, you could use results for gaseous Nitrous but lower the PROPEP input data temperature for this much colder gas. I decided that this was too much effort.]

PROPEP results: exhaust velocity

From Nitrous hybrids that I've built before, at the start of the burn I design the hybrid's combustion chamber pressure to be about 37 Bar (3,700,000 Pascals). The reason for this value I'll discuss later.

The PROPEP software says that our propellants burned at this chamber pressure and best mixture produce a mix of 21 different exhaust gasses. The ones occurring in quantity are nitrogen (N_2), carbon monoxide (CO), water (H_2O), and carbon dioxide (CO_2).

The exhaust is a mixture of gasses only; no hail of solid particles (such as soot) or other such toxic crud that you get with solid propellant motors (no condensed species).

The rocket is 21.3 kilometres high at the start of the burn, (launched from a balloon remember), giving a very low atmospheric back-pressure on the nozzle of only 4513 Pascals.

The nozzle's pressure drop (difference in pressure between one end and the other) is then

$$\Delta P = 3,700,000 - 4513 = 3,695,487 \; Pascals \quad \text{equ. 6. 10}$$

PROPEP says that this pressure drop will cause the temperature to drop. It starts at 3278 Kelvin in the combustion chamber and drops to 946 Kelvin as the gas leaves the nozzle exit.

This drop in temperature shows that the enthalpy has dropped as it transforms into kinetic energy. PROPEP says that 981 kilocalories (= 4,108,000 Joules) of enthalpy get converted into nozzle kinetic energy for every kilogram of propellants burnt in the combustion chamber. That's a *lot* of energy!

The equation for the specific kinetic energy of the gas leaving the nozzle is:

$$e_{kinetic} = \tfrac{1}{2} V_e^2 \quad \text{equ. 6. 11}$$

and this is equal to the specific enthalpy drop:

$$\Delta h = 4,108,000 \; Joules \quad \text{equ. 6. 12}$$

so:

$$\tfrac{1}{2} V_e^2 = 4,108,000 \quad \text{equ. 6. 13}$$

Rearranging this gives:

$$V_e = \sqrt{2 \times 4,108,000} = \sqrt{8,216,000} \quad \text{equ. 6. 14}$$

PROPEP works this out, and so can give us the exhaust velocity, which is

$$V_e = 2,866 \; metres \; per \; second \quad \text{equ. 6.15}$$

Notice how the exhaust velocity depends upon the square root of the change of enthalpy, and we know that the enthalpy depends upon the temperature of the gas in the combustion chamber. So the exhaust velocity depends upon the square root of the temperature of the gas in the combustion chamber. This is exactly the same result that we got in the last chapter, but using energy it was a lot easier to get.

PROPEP assumes we're using a perfect nozzle where the pressure thrust is zero, so this exhaust velocity is also the effective exhaust velocity. V_e = 2,866 gives a specific impulse (Isp) of $\frac{2,866}{9.81} = 292$ seconds.

Actually, this is the shifting flow result.

PROPEP also gives the frozen flow result, which is a bit lower because less enthalpy got converted to kinetic energy: 2,737 metres per second (giving an Isp of 279 seconds).

Isentropic flow

PROPEP had to make several assumptions about the gas flowing down the nozzle in order for it to be able to work out what the exhaust velocity was. These were the same assumptions a rocketeer would have to make if working the answer out using pen and paper.

Firstly, PROPEP assumed that the flow through the nozzle was adiabatic and reversible, so was isentropic (chapter 1). Then it could use a very helpful set of mathematical equations called the isentropic flow relations (see appendix 3) to let it work out the temperature drop ΔT down the nozzle just by knowing the pressure drop ΔP down the nozzle. Once PROPEP had the ΔT it could then work out the Δh, the change in enthalpy down the nozzle.

Is the nozzle flow really isentropic? Well it very nearly is, which is why the isentropic flow relations can give you very nearly the right answer for the actual nozzle temperatures, but not quite.

Energy losses

The flow isn't *quite* adiabatic: a few hundredths of the enthalpy escapes as heat through the walls of the combustion chamber and nozzle. And it's not quite reversible: as the gas slides along the wall of the nozzle it wastes energy by spinning up little whirlpools (friction).

So the real flow isn't completely reversible; you couldn't get all of the energy back, so really it's irreversible. But for our purposes it's close enough that we can treat it as being reversible.

There are other energy losses:

PROPEP assumed that all the chemical potential energy got converted into enthalpy in the combustion chamber, but actually we lose some of it.

The burning isn't perfect. In a real rocket, not all of the fuel mixed with all of the oxidiser, so not all of the propellants get burnt.

Also, not all of the gas coming out of our simple conical nozzle comes out directly straight. Most of it comes out the nozzle exit at a slight angle because of the angle of the exit cone.

This angle is then called the **nozzle divergence angle.**

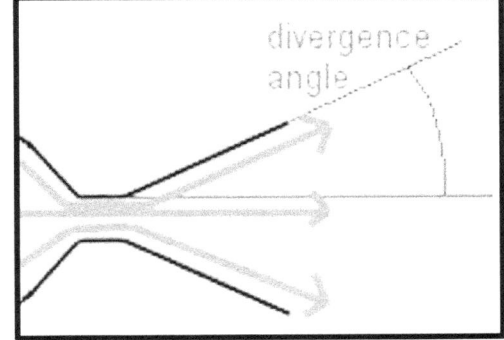

If you look at the gas coming down the centre of the nozzle, it comes out straight, so all of its velocity helps to make thrust.

But the gas coming out along the walls of the nozzle comes out at an angle: some of its velocity is sideways.

We can draw a vector diagram to find out the component of the sideways velocity that gives useful thrust (Valong), and how much is wasted on making sideways thrust (Vup):

You can see that Valong is a shorter line than the gas velocity line. This means that Valong is a bit slower than the gas velocity. So some thrust is lost.

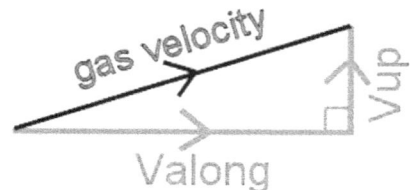

Notice from the picture of the Bell nozzle on page 62 that the nozzle wall is at a small angle to the long axis of the engine at the nozzle exit. The nozzle divergence angle of a Bell nozzle (see appendix 6) is low therefore little thrust is lost.

If the burning was perfect, and the nozzle flow was isentropic and didn't diverge, then the real effective exhaust velocity would be about halfway between the pessimistic frozen and the optimistic shifting PROPEP results for exhaust velocity.

Adding together all these losses, the effective exhaust velocity will actually be just a little bit less than the pessimistic frozen flow PROPEP results, so we'll use the frozen flow results from now on but multiplied by a loss factor of 98% (which I deduced from actual hybrid test-firings):

$$2,737 \times 0.98 = \underline{2,682 \text{ metres per second}} \qquad \text{equ. 6.16}$$

This effective exhaust velocity will give an I_{SP} of 273 seconds.

Post-combustion chamber

With hybrids, you can improve the mixing between the propellants by adding a **post-combustion chamber** between the end of the fuel grain and the nozzle. If they're better mixed, they burn more completely.

This is just a length of empty combustion chamber that is about one chamber-diameter long:

The exhaust gasses pouring out of the fuel port flow over the end of the port and round the post-combustion chamber in a doughnut shaped swirl (called a toroidal vortex).

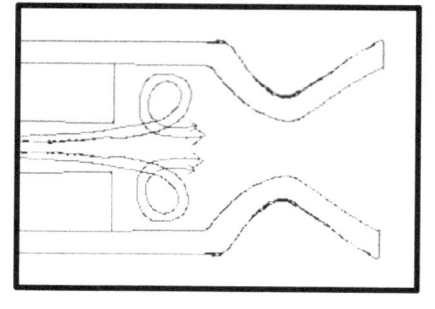

The propellants get thoroughly mixed during this swirling and any unburnt propellants get the chance to burn. Adding a post-combustion chamber actually makes a big difference to the performance of hybrids, well worth the extra chamber wall mass it causes.

We'll add a post-combustion chamber to our hybrid; the above I_{SP} of 273 seconds assumes that we've added one.

[Note that the above PROPEP results are for liquid Nitrous entering the combustion chamber at 293 Kelvin (20 degrees C). As the burn progresses, the Nitrous in the tank will get progressively colder: the loss factor of 0.98 above is based on the average Nitrous temperature during the burn.]

The nozzle

PROPEP assumed we were using a perfect nozzle. That is, one that had just the right area ratio so that the pressure of the gas at the nozzle exit was exactly equal to the atmosphere's back-pressure at 21.3 kilometres up.

This gave zero pressure thrust, so the effective exhaust velocity was equal to the real exhaust velocity. PROPEP says this perfect nozzle required an area ratio of 50.

We'll use a bigger nozzle because we're going on to higher altitudes. (This makes the nozzle larger than it needs to be at the start of the burn.)

At these high altitudes, we can use a wide nozzle with a larger area ratio to almost get the exhaust gasses down to the extremely low pressure outside at the edge of Space. We could be tempted to double the ratio up to 100, but the nozzle might be heavy, so prudence dictates an area ratio of 70.

Actually, this nozzle will be too big at the start. What happens then is that the exhaust gas tries to follow the nozzle as it gets ever wider, but it can only do it for so long. Then it just gives up and separates from the nozzle wall. This is called **separated flow**.

American rocketeers get very worried that flow separation will bend their nozzles out of shape but British rocketeers don't worry, they just make their nozzles stronger.

Separation causes a slight loss of thrust but not much. We'll ignore the loss, because we'll only get separated flow for the first few seconds of the burn.

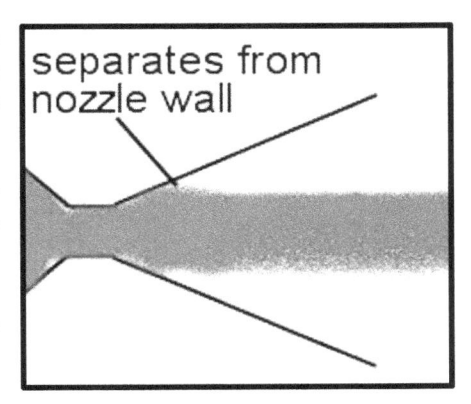

Chapter 7: The hybrid combustion chamber

In a hybrid engine, the combustion takes place just off the surface of the solid fuel, between fuel that has just been vaporized by the heat of combustion, and oxidiser injected into the head of the combustion chamber.

The Boundary Layer

In a hybrid, the combustion takes place within the **boundary layer** of flow affixed to the inner wall of the central port.

The flow of fluid down a pipe or a tubular fuel grain port can be *imagined* as split into two separate regions:

In the central region, the fluid is all flowing at the same flow speed, but in a concentric outer region, the flow speed drops with distance above the wall of the port, coming to rest as it touches the wall.

In reality, the flow speed is different right across the port - with the maximum in the centre - but it reaches 99% of the central flow speed only a few millimetres from the wall, and that is defined as the boundary between the regions.

This 'boundary layer' is - and appears to be - stuck to the inner port wall. It sticks there because the surface of the fuel grain is actually very rough and craggy at microscopic scale, and effectively 'snags' the layer of flow next to it.

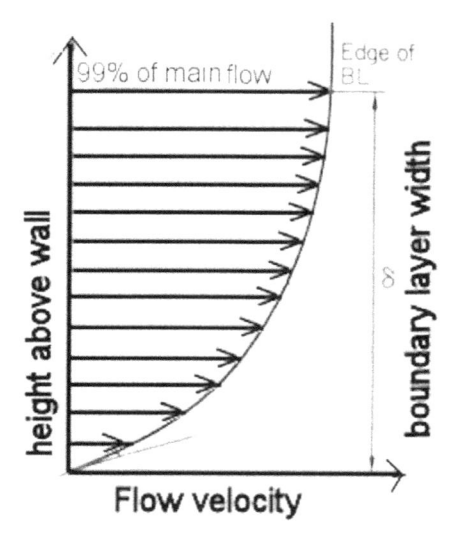

You can visualise a radial slice through the boundary layer in a very wide pipe as a pack of cards lying on a table, where successive cards in the deck represent successive plane, parallel layers of fluid at increasing distance from the wall of the port (from the surface of the table).

Drop the pack onto a table but with a sideways movement. Successively lower cards in the pile move forwards with lower velocity; the lowest card remains stationary on the table. This is exactly how the layers of fluid within the boundary layer behave.

Growth and turbulence

This ordered mental picture of parallel laminate layers, known as **Laminar flow** needs to be modified somewhat:

For one thing, each layer gets thicker as it flows down the pipe, as if each card was thicker at the downstream end than at the other. This is termed **boundary layer growth**. This is due to the entire boundary layer slowing down with increasing length, so an ever wider region of flow is slower than the central port flow.

Also, at some point along the length of the port, the boundary layer breaks up; what determines this 'transition' to chaotic, or **turbulent** behavior isn't well known.

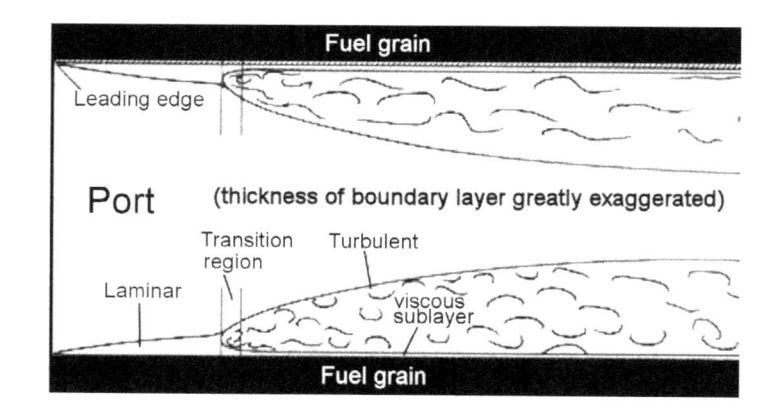

A turbulent boundary layer is much thicker than a laminar boundary layer, and can be thought of as a fractal series of many little eddies within eddies, the separate layers of fluid within the boundary layer get thoroughly mixed together.

If the flow in a laminar boundary layer hits a step or cliff, it will trip to turbulent flow, though the boundary layer will 'trip' turbulent all by itself simply after having had to flow over the walls for a long enough period of time.

You can see this effect clearly in the smoke rising from a cigarette: the smoke is initially laminar, but after a critical distance above the cigarette it suddenly goes turbulent, breaking up into thick eddies. This distance is typically a few centimeters down a hybrid port.

The fuel grain port boundary layer

Once the hybrid has been ignited, some of the hot combusted gas is carried via the turbulence within the boundary layer to the surface of the solid fuel where its heat is conducted into the top layer of this fuel (a process known as **forced convection**; radiation plays a minor role).

The heat then melts a layer of fuel off of the exposed surface of the fuel charge. 'Melt' isn't quite the right phrase with plastic fuel grains, because the long polymerised hydrocarbon molecules in plastics are too long and entangled ever to have solidified completely at low temperature like the nice, ordered atoms of a solid metal: they never really froze. Instead, they're supercooled liquids with phenomenally high viscosity as with glass, but once heated above a critical **Glass Transition Temperature**, they will flow like a liquid, albeit treacle.

As the surface is heated further, the remaining liquid vaporizes to gas. The fuel may char or slough off tiny chunks before melting, and then it de-polymerises: the carbon-to-carbon bonds holding the backbone of the polymers together absorb too much thermal energy and break apart, and the fuel turns into a mixture of several gasses and simpler hydrocarbons.

With further heating, the cracked hydrocarbons vaporize to gas and get carried off into the main flow. This is the primary transport mechanism that gets the fuel and oxidiser mixed.

Once transported into the port flow, the vaporized fuel burns in a narrow zone within the boundary layer known as the **flame zone** or 'flame sheet' [a 'macro diffusion flame'], which is at a depth between the fuel grain below and the oxidiser flow above that is close enough to stochiometric to support combustion.

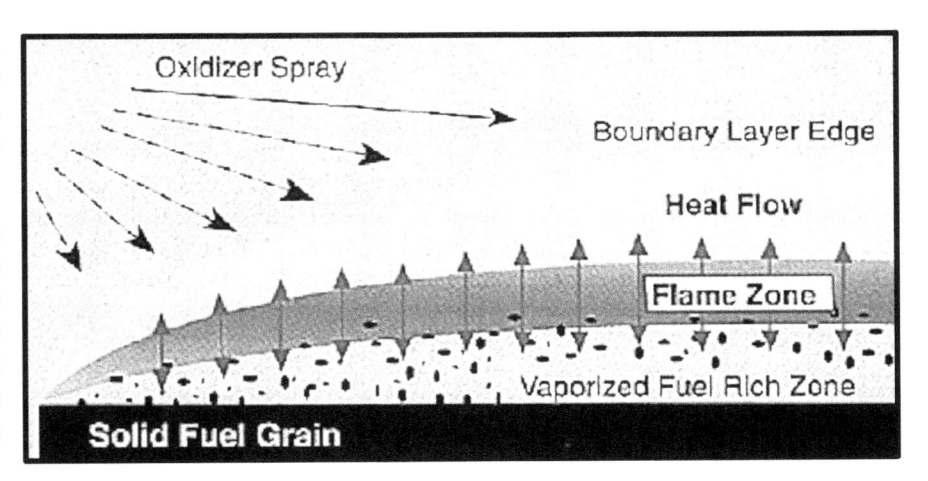

Regression rate

The **regression rate** $\dot{r}$ is the speed (typically millimetres per second) that the solid fuel grain port surface is eroding.

This rate at some point x metres down the port of the hybrid is typically a function of the mass flux G (see chapter 1) of gas that the grain is suffering at that point along the port. G is simply the total of the fuel mass flux plus the oxidiser mass flux:

$$G_{total}(x) = G_{fuel}(x) + G_{ox} \quad \text{equ. 7.1}$$

Thus, in a hybrid, the fuel vaporized depends on the mass flux of oxidiser flowing down the port, and so maximizing G_{ox} by increasing its mass flow rate (by opening up the injector orifices) - and also by decreasing the port diameter - will maximise thrust.

If you can force the mass flow rate to flow down a narrow port then the mass continuity equation (equation 1.18) says that the flow's velocity v will have to increase as the area A_{port} decreases, to get the same mass flow rate through.

A faster flow can grind more fuel off the wall of the port (think of sand-blasting). So the regression rate $\dot{r}$ depends on the mass flow rate *and* the width of the port i.e. the mass flux.

There is a limit, however, to the maximum mass flux attainable within the port. If too high, then the flame will literally be blown out. The maximum oxidiser flow rate occurs on start-up because the chamber is still at low pressure, and so the pressure difference between tank and chamber is at maximum.

The more energetic the ignition system, the higher flux the engine can withstand without flameout. Also, site the ignition system just downstream of the injector so that it decomposes the Nitrous to release hot oxygen that then meets the fuel grain.

Regression rate equation

A typical equation for the regression rate $\dot{r}(x)$ of a plastic fuel burning with oxidiser is:

$$\dot{r}(x) = a\, G^n(x)\, x^m \quad \text{equ. 7.2} \quad \text{(in metres per second)}$$

where a, n, and m are constants dependent on the propellant combination, and x is the distance down the port.

The mass flow rate of fuel vaporized within the port - per square meter of port surface area - at the x of interest is then:

$$\dot{m}_{fuel\_per\_unit\_area}(x) = \rho_{fuel}\, \dot{r}(x) \quad \text{equ. 7.3}$$

and this is integrated along the length of the port to get the total mass flow rate of fuel heading for the nozzle:

$$\dot{m}_{fuel} = \int_0^x \rho_{fuel}\, \dot{r}(x)\, S(x)\, dx \quad \text{equ. 7.4}$$

where $S(x)$ is the port surface area at point x.

The above equation for $\dot{r}(x)$ suggests that the diameter of the port will increase (regress) much more rapidly at the downstream end (larger x) than the upstream end. However, the boundary layer (whose thickness partially blankets the fuel grain from the heat of combustion) grows larger with x as well, so the diameter of the port grows only slightly larger along its length.

Furthermore, the burning is self-levelling: any 'humps' in the fuel grain increase the local erosion rate around themselves to smooth themselves out.

For these reasons, much more useful expressions for comparison with test-firing results can be based on the average values along the port at some time point of interest during the burn:

$$\dot{r}_{average} = a\, G_{average}^n\, L^m \quad \text{equ. 7.5}$$

where x has now been replaced by L, the total port length, and:

$$G_{average} = \frac{\dot{m}_{total}}{A_{average}} \quad \text{equ. 7.6}$$

i.e. based on an average measurement of the port cross-sectional area.

The values of a, n and m are now (slightly) different (values of a are always jealously guarded.)

The mass flow rate of fuel vaporized by the end of the port is then:

$$\dot{m}_{fuel} = \rho_{fuel}\, \dot{r}_{average}\, S \quad \text{equ. 7.7}$$

where S is now the current total port surface area $\pi\, D\, L$ (assuming a circular port, otherwise use the **hydraulic diameter**).

[For a port of non-circular cross-section, we need an equivalent to the diameter of a circular port in order to calculate mass flux. This equivalent is known as the hydraulic diameter and is defined as:

$$D_H = \frac{4A}{P} \quad \text{equ. 7.8}$$

where A is the cross-sectional area and P is the perimeter of the cross-section.]

Note that $\dot{m}_{total} = \dot{m}_{fuel} + \dot{m}_{oxidiser}$ is the *total* mass flow rate at the end of the port and hence flowing towards the nozzle. Therefore, the solid erosion is a cascading effect: the more fuel gets eroded, the larger the total mass flow rate, hence the more fuel that gets eroded. It's a feedback mechanism, but fortunately it converges to a final answer. Needless to say though, an implicit, iterating software loop is therefore required to get the final value of $\dot{m}_{total}$

Excel spreadsheets will handle iterative calculation loops well, though you've got to tick the iteration function to on. (On the top menu select: file [or 'tools' on older versions], options, formula, calculation options, and tick the box 'enable iterative calculations'.)

A simple, robust, and rapidly converging algorithm you can use for this is:

1. Start by assuming that $\dot{m}_{total} = \dot{m}_{oxidiser}$ only (i.e. $\dot{m}_{fuel} = 0$)

2. Determine $\dot{r}_{average}$ and hence $\dot{m}_{fuel}$

3. Add $\dot{m}_{fuel} + \dot{m}_{oxidiser}$ to get a new $\dot{m}_{total}$

4. Has the new $\dot{m}_{total}$ settled down (converged) to equal the previous step's $\dot{m}_{total}$ to the required accuracy? If it hasn't, use this new $\dot{m}_{total}$ and loop back to 2. If it has, the calculation is complete.

Excel handles the final check automatically: you simply set the convergence limits in the same sub-menu, and then make spreadsheet cell 1 equal to spreadsheet cell 3 to close the loop.

The exponent m in the above regression equations tends to be a very weak power of length, much less than 1.0, so for the moderate port length changes involved in tuning a hybrid for performance, it can be ignored.

Furthermore, when modifying an existing design, one will use the same ignition system, so will expect to see similar fluxes down the port. In this case, you can get away with an engineer's simplification of the regression equation:

$$\dot{r}_{average} = a\, G_{liquid}^n \quad \text{equ. 7.9}$$

where the length term has been lumped-in to the catch-all coefficient a.

This is much easier to use as it's based only on the liquid flowing in from the injector, so no iterative calculation is required. This equation simply won't work if you double the size of the hybrid for example, and may well not be transferable to someone else's design.

Appendix 5 details how to convert regression rate data between different systems of units.

Typical values

The trouble with plastic fuels is that their mechanical strength means that they're quite resistant to being ground down by the hot gas flowing past them inside the port of a hybrid. This means that their regression rate is low.

We at Aspirespace have spent considerable time and money performing test-firings to find the regression rate equation for our hybrid fuels, so we don't give this information out.

However, for the typical port mass fluxes you'll want to use, then as the mass flux drops throughout the burn, the following table of average regression rates will be a good guide for HDPE and similar plastics burning in Nitrous oxide:

Regression rate mm/s	Liquid burn	Vapour burn	Average-entire burn
Small HPR hybrid	1.2	0.3	0.8
Large hybrid	0.8	0.2	0.6

The difference with engine size reflects the term in the regression rate equation based on port length.

Note that the data for large (man-carrying) hybrids has been extrapolated from small hybrids and so is a rough estimate. We at Aspirespace haven't fired such a large hybrid as yet.

Maximum flux

Design the port diameter to suffer a maximum mass flux of 870 kilograms per second per square metre as the engine comes up to operating combustion chamber pressure, as a higher mass flux can blow the flame out. This value is based on a powerful pyrotechnic ignition system (a lump of solid propellant just downstream of the injector) and may blow out weaker ignition systems.

Swirl: vortex hybrids

Implicit in the regression analyses above was that the mass flux was uniform across the port, but it needn't be. Using purposefully angled injector orifices, a vortex (swirl) can be established along the port. Adding high rotation to the port flow centrifugally compresses the flow far from the central axis (increases its pressure and density). So the local mass flux at the burning surface is considerably higher than the overall mass flux averaged across the entire port. Thus it appears that the regression rate has been increased, perhaps threefold, but for the same averaged port mass flux. Obviously, it's much easier to maintain a vortex down the port if the port has a circular cross-section.

Fuel selection

The thrust at any point in the burn depends primarily upon $\dot{m}_{total}$ - the total mass flow rate of fuel plus oxidiser within the port - which itself depends upon how rapidly the solid grain is regressing; so the higher the thrust required, the more rapidly the port diameter has to increase.

This poses a geometrical problem:

Taking the simplest case of a fuel grain with a single port of circular cross-section, the cross-sectional area

$$A = \pi r^2 \quad \text{equ. 7. 10}$$

so from the chain rule of differentiation, as the port regresses, this area increases with time at an over-linear rate:

$$\frac{dA}{dt} = \frac{dA}{dr}\frac{dr}{dt} = 2\pi r \dot{r} \quad \text{equ. 7. 11}$$

Unfortunately, assuming a constant oxidiser mass flow rate into the port, this means that the total port mass flux is decreasing at this over-linear rate because (from the mass flow rate formula) it is inversely proportional (=1/A) to area A. So the thrust will drop, and so will the solid propellant mass flux, which means that the burn is getting progressively leaner (too little fuel) with time. So even more thrust is lost.

Fortunately, as the port widens, its total surface area S (= $2 \pi r L$ which is the area available to farm fuel mass flow from) increases, but it's only increasing linearly with the regression rate:

$$\frac{dS}{dt} = \frac{dS}{dr}\frac{dr}{dt} = 2\pi L \dot{r} \quad \text{equ. 7. 12} \quad \text{(where } L \text{ is the total port length)}$$

whereas as we've just seen, the mass flux is dropping over-linearly, and so is dominating.

So contrary to popular opinion, the best hybrid propellant combination is the one that has the *lowest* regression rate that you can use for your application, because the thrust and fuel-to-oxidiser ratio will then wander the least with time.

A low regression rate tends to require a long port to provide enough surface area to farm sufficient solid propellant, which creates a long combustion chamber of thin diameter.

Scaling up a hybrid

If designing a fuel grain for a small HPR rocket vehicle, you would design its fineness ratio (its length compared to its diameter) based on diagrams you would have seen of similar small hybrids.

But they're only around half a metre long. When you double the size of a hybrid, the length of the fuel grain becomes two times longer. But the cross-sectional area of the fuel grain's port becomes 4 times wider (= 2 squared), and the mass and volume of the fuel grain become 8 times larger (= 2 cubed). Such is the way that volumes and areas grow disproportionately large as lengths grow.

All these things are growing by different amounts, so a large hybrid's fuel grain needs to be a completely different shape to a small hybrid's fuel grain.

As you make the fuel grain larger, the surface area of the port grows less rapidly than the fuel mass does, so pretty soon there isn't enough surface area to burn all of the fuel off of. So you need to provide much more burning area whilst avoiding an inordinately long fuel port; the way to do that is to have more than one port (called a **multiport hybrid**).

There are two ways to do this: Scaled Composites' SpaceshipOne used a hybrid fuel grain with four pie-shaped ports in it:

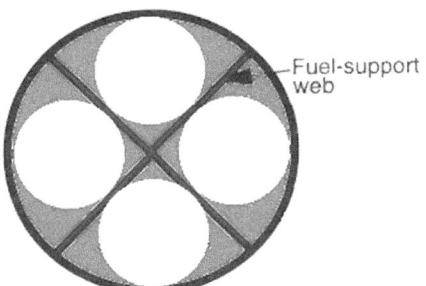

The ports have to be slice-of-pie-shaped as shown above because circular ports leave too much unburnt plastic remaining at burnout as shown here:

But getting HDPE moulded into pie shapes is a considerable expense, and you might need to change the size later in the development programme which will need another mould to be made. [Fortunately, experimenters have recently found ways to use recycled HDPE in 3D printers which lets you print any shape of port that you like once you build a suitably large 3D printer.]

For this reason, Scaled Composite's SpaceShipOne and SpaceShipTwo use a Hydroxyl-Terminated PolyButadiene (HTPB) fuel which is easily moulded.

There's another way to get a multiport hybrid that can use cheap and simple circular tubes of HDPE, and that's to have more than one combustion chamber, but with the multiple chambers exiting into one common nozzle. This is the system we will use.

Ignition systems

I've included the ignition system within this chapter, because it resides within, or acts within, the combustion chamber.

The idea is to supply enough heat energy to the fuel grain so that it will gasify and ignite in the presence of the oxidiser.

If the oxidiser is oxygen, then ignition is trivial, as the ignition temperature of hydrocarbons (and much else) in oxygen is low. With Nitrous oxide however, the fuel grain has to be hot enough to decompose the Nitrous on contact, which requires a high temperature.

If the ignition system can also continue supplying heat for the first few fractions of a second after the Nitrous has begun to flow into the combustion chamber then so much the better, as it will decompose some of the Nitrous *before* it reaches the fuel grain. For this to happen, the ignition system has to reside/act at the top of the engine next to the injector.

Some hybrid ignition systems use a gas torch to light the fuel grain. This can be built into the forward engine bulkhead, or more often - to save weight - is part of the launchpad, and is in the form of a long, thin pipe that enters through the nozzle and continues up the fuel port.

Pyrotechnic systems

The ignition method that is by far the best - based on our experience - is a pyrotechnic system. Solid propellant ignition has the highest energy density (heat delivered per system mass) of ignition systems, and is generally the most reliable for amateur experimenters, provided a decent commercial electrical igniter is used. (An igniter is a coil of wire that glows red-hot due to resistance when electricity flows through it.)

A pyrotechnic system will work thusly:

You fire the igniter by remote control. This lights the **heater charge**: this is a cylinder of solid propellant that burns, spewing hot lumps of solid propellant down the inside port of the hybrid's tube of fuel. This heats the fuel, ready to receive the Nitrous.

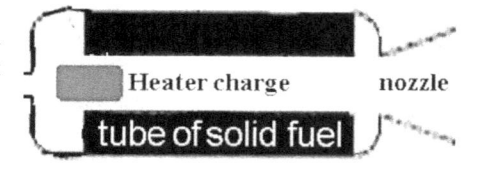

Then you remotely open the run-valve between the Nitrous tank and the combustion chamber. Once the valve is open, the Nitrous flows through the injector into the combustion chamber. It hits the burning heater charge and decomposes into nitrogen and oxygen. The oxygen hits the hot plastic fuel tube and the tube port bursts into flame.

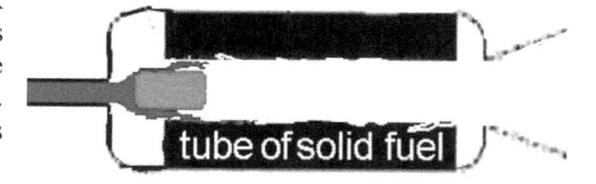

For safety, make sure the heater charge can't break free and block the nozzle, and for added safety, make it long and thin, so that if it *does* break free it can pass safely through the nozzle throat and out.

As solid propellant weighs little, don't scrimp on it; one can never use too much of it for ignition, though don't turn your combustion chamber into a solid motor!

Pyrovalve

At Aspirespace we use a **pyrovalve** system to ignite our HPR-sized hybrids. A pyrovalve is a suitably robust plug of ultra-dense solid propellant that blocks the output of the injector, so plugs the flow of Nitrous. When lit, the pyrovalve not only disintegrates to let the oxidiser through into the combustion chamber, but also pyrotechnically ignites both the Nitrous and the fuel grain port as a heater charge.

U/C valve

One pyrotechnic system popular with commercial and hobbyist HPR hybrids is the so-called U/C 'valve' (**Urbanski-Colburn valve**) wherein a plastic run-tank fill line is physically attached to the engine upper bulkhead via a pneumatic fitting which becomes the injector. This fill line runs up from the fill-tank through the nozzle to the upper bulkhead.

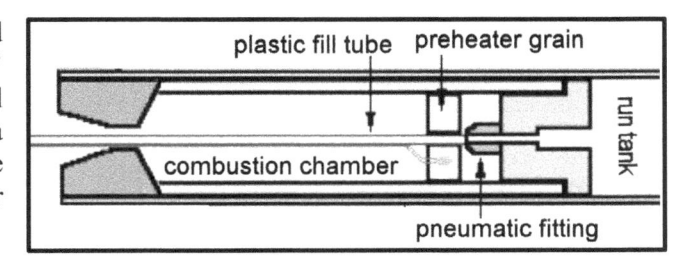

When the engine lights, this plastic tube melts through, releasing the rocket vehicle from its ground support.

The burn-through plastic fill tube is an interesting idea that can be expanded upon, but in my opinion (no disrespect to Urbanski or Colburn), the U/C valve causes more problems than it solves and is best used only on very small hybrids. As well as giving an undefined launch time, the lone pneumatic fitting makes a very poor injector.

Some large commercial HPR hybrids use several U/C fill lines in parallel (therefore several pneumatic fitting orifices) which will not all burn through at the same time. Frankly, this is a terrible idea and is best avoided: there's no substitute for a proper injector with many more orifices.

High altitude ignition problems

It would be embarrassing if you got your spacecraft to very high altitude on a balloon but then the rocket failed to ignite.

All of your static test-firings of the engine, (and any parachute deployment systems), will be done at ground-level. Below about 20,000 feet altitude, air contributes significantly to the heat transfer from the igniter to the rocket propellant or parachute expulsion charge. Above that altitude there is significantly lower heat transfer, so a much more energetic igniter is required to set off the propellant than at ground-level.

Several HPR rocket vehicles have suffered parachute failures at very high altitude: the expulsion charge of powder didn't ignite. It's not clear what went wrong in these cases but it's thought that the near-vacuum of very high altitudes prevented the propagation of heat/flame across the loose pile of **expulsion powder**: the bulk of the powder didn't burn.

[Expulsion powder - which can be Black Powder or an equivalent - is effectively gunpowder, a loose pile (a 'charge') of pyrotechnic powder designed to expel a parachute out of the barrel of effectively a cannon.]

Expulsion powder, and also the solid propellant of the hybrid heater charge, has a **deflagration limit** which is a minimum pressure at which combustion is barely self-sustaining. If the pressure is too low, combustion will cease or be erratic at best.

One way to solve this problem is used on military and commercial high altitude rocket vehicles: they use sealed canisters containing sea-level pressure air to contain the powder or propellant. The container is designed to burst at a set pressure using a burst disc once the powder/propellant has ignited properly and the ensuing combustion gas expands. Whatever material is chosen for the burst disc should be tested to make sure it will break at a 1.5 Bar overpressure to prevent fragment damage to the rocket vehicle, since confined expulsion powder can generate 1700 Bar pressure or higher.

[A **burst disc** is a thin blanking plate across a pipe that is scored across its face. It bursts at a set pre-determined pressure. Burst discs can be bought from many vendors.]

For our high altitude hybrid engine, you would seal the nozzle with a burst disc to trap sea-level air inside the combustion chamber. Use a plastic or aluminium disc so that it melts away completely once the engine is firing. This disc is often glued across the throat to keep the physical size of the disc small.

Chapter 8: Nitrous hybrid safety

Nitrous hybrids are one of the safest forms of rocket engine, hence their popularity with amateur researchers. However, any system of stored energy (chemical, high pressure) has the potential to fail catastrophically, and people can be injured or even killed unless safety precautions are taken.

Hazards analysis

The first step is to design safety into the system from the start. Perform a hazards analysis on the engine system, particularly the oxidiser components:

- Determine the most severe operating conditions (pressure, temperature, flow velocity, rubbing of components).

- Evaluate the flammability of the engine plumbing materials when in use. Test unusual materials for flammability.

- Evaluate potential unwanted ignition sources: flammable materials can only be used in sections of the plumbing where there is absolutely no chance of ignition.

- Be cautious of **single-barrier failures**: a single-barrier failure is a leak in which only the primary containment structure is breached. Such a leak introduces oxidiser into a region not normally exposed to oxidiser. The materials or configuration of parts in this region may not be compatible with oxidiser! Single-barrier failures are often overlooked, but the potential for component-part failures, such as diaphragm failures, can cause hazardous oxidiser-enriched environments, and can cause a substantially increased risk of ignition near electrical components for example.

- Attempting to correct single point failures simply through procedural actions is not a reliable method. Design the system safe instead. [A single point (of) failure is a part of a system that, if it fails, will stop the entire system from working. These are undesirable in any system with the goal of reliability.]

- Include hazards associated with contamination (muck, dust).

- Review secondary hazards, such as seal leakage to electrical equipment.

- Design equipment so that power losses, control pressure leakage, or other loss of actuation sources, return the equipment to a fail-safe position to protect people and property in the event of an accident.

Nitrous safety

Nitrous is a powerful anaesthetic and is used as such by the medical profession: inhale too much and you might never wake up again. You can carry bottles of Nitrous in the back of vans but not in cars; there should be a barrier between driver and bottle in case of a leak because of the powerful anaesthetic effect.

Storing/filling oxidisers

Store fill-tanks out in the open, and keep all their plumbing above ground (no buried pipes) to aid leak detection. Have a source of water nearby for firefighting/deluge/dilution.

Cleanliness

…is next to godliness as they say. For oxidiser systems this is imperative, as most contaminants are at the very least fuels, and often will cause an explosive compound with oxidisers. Use special degreased tools, and only use those tools for the oxidiser system alone. Use lint-free cloths, and avoid creating metal particles/swarf. Degreasing the plumbing is dealt with later in this chapter.

Training

The safe use of rocketry oxidisers is all about education. The rocket firing team must all be trained to handle the oxidiser being used, understand its chemistry, its materials compatibility and non-compatibilities, and be aware of all safety implications.

Fire

One of the main hazards of oxidisers is their ability to create a fire. All three sides of the triangle must be present: a fuel, the oxidiser, and an ignition source. But with rocketry oxidisers, the fuel could even be metal pipework, and the ignition source a tiny spark from static electricity, or friction.

The key to the proper use of rocketry oxidisers is the use of compatible materials and the elimination of ignition hazards.

Pressure vessel and plumbing mechanical safety

Manual versus remote fill as it affects propellant tanks

Any container that can safely withstand the pressure without bursting will do for a propellant tank. This can be filled manually, or remotely from a distance through automation which is preferred for safety. A manual fill system is simpler to construct, but involves a higher level of design and testing of all tanks and plumbing that will be pressurised; safety of the filling crew is the prime consideration.

Be aware that a vessel pressurised full of high-pressure gas or vapour is a potential grenade. The United Kingdom Rocketry Association (www.ukra.org.uk) rightly regards standing anywhere near a cylinder under pressure as potentially hazardous. Their safety code only allows people to be close enough for them to manually fill a tank if the tank is deliberately designed with a high margin of safety to prevent bursting. For experimental hybrids UKRA will rightly require sufficient proof of tests performed to verify the tank safety if anyone is to be near it.

Remote-filled systems allow the luxury of smaller tank safety margins, and so thinner-walled, lighter tanks, but the potential mass saving if using an ultra-thin tank is proportionately smaller as the vehicle diameter gets smaller. For example, small HPR aluminium tanks suffer minimum gauge problems, and over-proportionally large bolt-holes which raise stress.

If nobody's *ever* going to be near a tank while it's pressurised, i.e. only remote filling and dumping throughout the whole life of the tank, then the safety-factor on burst can be low to keep the tank mass down. 1.5 times the expected pressure is a reasonable remote-fill margin for a reusable tank, but pressurise the tank and plumbing a few times to this pressure before first flight to be sure.

Obviously, you *can* safely be close to the massively over-strengthened fill-tanks used to transport Nitrous from the suppliers; they're properly engineered and manufactured to allow you to move the contents around the country after all, and they're tough unless you drop them on their necks and break the valve off. Then you get a rocket alright: the tanks can fly at very high speed and cause major damage. Such a commercial container is safe enough to handle (though not idiot-proof) but is far too heavy to fly.

Thin-wall aluminium tanks

How strong do propellant tanks have to be to be *reasonably* safe to be near without being too heavy to launch? Military missile-design literature recommends a design safety-factor of *at least* 2.0 times expected internal pressure on pressure vessels that are to be manually handled while they're pressurised.

But this applies to a *production* tank; the point is that a batch of military production tanks have been tested to destruction to ensure that they really do burst at well over the 2.0 safety-factor, and not a worrying 2.00001. Amateur groups don't tend to make batches of tanks, therefore can't perform a suitable number of destruct tests.

Several rocketeers making one-off tanks, have learnt to their cost that home-made thin-walled tanks, especially aluminium ones, need a burst safety-factor of at least *three* to handle the inevitable dings, scratches, and accidental alteration of the metal properties (typically inadvertent annealing) during manufacture. Unexpected stress raisers can cause aluminium to rip apart at much lower *average* stresses than you bargained for.

So for manual filling, then Aspirespace's approach, based on the personal experiences and advice from engineers who make home-made tanks, is that you should design your hybrid's one-off tanks to ensure that they safely withstand three times the **Maximum Expected Operating Pressure** (MEOP) without **yielding** (see chapter 10). For Nitrous oxide this would be three times the vapour pressure of Nitrous on a hot day, or 3 times 60 Bar = 180 Bar.

You should design the tank to fail in the safest way possible: popping the end-caps off rather than bursting the wall.

Having designed the tank to three times MEOP, you would then 'proof' test the tank to twice MEOP (a hydro-test: see later in this chapter).

Home-made aluminium thin-wall tanks are often formed from end-domes and cylinders welded together. The 6000 series aluminiums are usually heat-tempered (designation 'T') and when welded lose this temper and so lose considerable strength. The tank therefore needs a post-weld heat treatment to restore the temper. The 2000 series aluminiums, though weaker, are not tempered so avoid this problem. Another way to get around this problem is to add extra metal around the joint: this lowers the stress in the metal so that the local loss of strength isn't an issue. Bolting the end-caps on avoids the welding issue.

Composite tanks

Several decades ago, Space and aerospace companies identified the need for low mass pressure vessels. The composite overwrapped pressure vessel was developed to provide weight savings without a loss in safety margins over traditional tanks. The design consists of a thin liner of material, typically a metal, overwrapped with a continuous fibre yarn impregnated with a resin such as epoxy. The weight saving over a traditional tank is around 25 to 50 percent.

The general advice from composite manufacturers is that home-made composite tanks have resulting properties that are far too variable from tank to tank to predict: *use remote filling only.* Or buy a commercial tank. More to the point, when composites are damaged or experience many pressurisation/depressurisation cycles, the ensuing micro-cracks in the composite are rarely visible on the surface but will grow under mechanical stress, so the tank may rupture the next time it's pressurised without any visible prior warning.

Even thin-wall aluminium liners are not thought to provide the whole structure with enough uniformity to allow anyone to be near a home-made one when pressurised.

One would hope - and industry demands - that the tank would leak before bursting, releasing the pressure. But this confidence only comes from very carefully controlled manufacturing, so that any cracks are known to be 'stable'. Stable can be defined as a crack that has a length less than 10 times the wall thickness of the vessel, and has a stress intensity less than the material toughness.

Don't assume that a leaking liner makes the whole tank safe; a failed liner transfers more of the load to the composite, which can then undergo brittle failure. Make the composite strong enough to carry the entire pressure load.

Composite tanks need a separate internal liner because raw composite leaks the contents inside through microcracks, and furthermore the resin can cause Nitrous oxide to 'explosively' decompose, or form an explosive with Lox and probably High Test Peroxide.

Plastic liners are lightweight, but you must ensure that the plastic is one that isn't a fuel or catalyst: use the fluoroplastics such as PTFE or FEP.

Composite tanks are very sensitive to impact: write an impact-mitigation work plan (e.g. no big metal tools to be dropped onto the tank) and consider safety shields around the tank.

Tank testing

No pressure-vessel, no matter who or what company made it, is at all safe until it's been pressure-tested. Until somebody's tested it, *don't use it*. Don't even assume that a tank specifically sold for HPR Nitrous hybrids will take full Nitrous pressure; some are only designed to take the lower pressure that you get with a permanently-open large-vent-hole system.

Hydro-testing: a safer testing method

Before you fire your hybrid for the first time, you first have to make sure that its tank, its combustion chamber, and all the plumbing between the two, will safely withstand the high pressure they're expected to encounter without bursting. To do this you need to remove the nozzle and then fit something to close the hole off to seal the chamber for testing.

The dangerous way to test the rocket (so *don't* do this) is to fill it with compressed gas at high pressure.

This is dangerous because if the tank or chamber bursts, the compressed gas inside is let loose, and will expand to many times its original volume. This rapidly expanding gas will throw bits of broken rocket at several hundred miles an hour, causing as much damage as an explosion.

The *safer* way to pressure-test the rocket is called a hydro-test (a hydraulic test). This involves pressurising the rocket with a liquid instead of a gas.

Pieces of burst tank or chamber won't go very far under liquid pressure, because liquids can't store energy in compression nearly so well as gasses.

You can either use water as the test liquid, or use hydraulic jack oil which is less likely to rust the pressurising pump.

Here's a picture of Aspirespace's Adam carefully filling a hydraulic hand-pump full of oil, ready to do a hydro-test:

He's secured the barrel of the pump facing upwards to make sure he gets every last bubble out of the pump. There must be absolutely no bubbles of air in the pump or rocket during the hydro-test.

You should do the hydro-test remotely from a safe distance, preferably behind a wall.

If you do use jack oil, then once the hydro-testing is finished, clean the inside of the tank and plumbing because jack oil is a fuel, and it could burn explosively with the Nitrous oxidiser.

Hydro-test the rocket up to higher pressure than it will experience during firing. If you used a safety-factor of two, then hydro-test it up to 1.5 times the expected pressure.

The proof test

You're looking for proof that the tank won't rupture; hydraulically test the tank to 1.3 or 1.5 times the highest pressure you ever expect the system to encounter (MEOP). The proof test isn't a to-destruction test, so the safety-factor is lower.

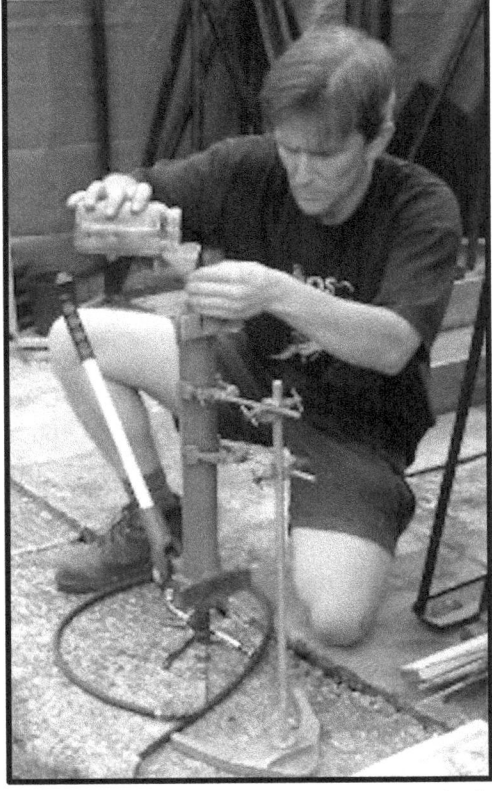

Leak testing

Once the system has been successfully hydro-tested and you know it can take high pressure, *only then* can you safely make a *low*-pressure leak test using compressed air or nitrogen (6-10 bar usually works fine: you can use a car foot-pump). If using air from a compressor, make sure it doesn't contain oil.

A leak test can save you much expensive Nitrous if you would otherwise have to dump a full run-tank to fix a leak.

You can buy "snoop" leak detector, which is basically washing up liquid in water: where there's a leak, bubbles froth-up which you can see easily.

Water

After testing is over, it's important to make sure that all water in the pipework is removed as it can cause valves to freeze stuck. (Nitrous can fall below water freezing temperature when it expands). Water will hide inside cavities such as the internal passages of valves. You need to blow it out with compressed air or nitrogen, or cook it out at 50 degrees C.

Hydraulic overpressure, the head space

Pick up any fresh bottle of camping-stove-gas or CO_2 and give it a gentle shake; the sound of waves sloshing inside reveals that the bottle hasn't been completely filled with liquid, there's obviously a small amount of gas in there as well.

This small 'head space' or 'ullage' of gas is in there for a purpose: because after filling, the liquid's density will change with any future changes in temperature. In fact just-subcritical fluids like Nitrous, carbon dioxide, or camping gas, can change density rather a lot with temperature if they're around room temperature.

The danger is that if the temperature increases, the liquid density will drop (see chapter 2). If the tank was completely full of liquid, then the tank's fixed volume now won't be enough to contain the mass of liquid as it expands (i.e. increases its volume). If the tank is stoppered, the liquid will then self-pressurise. Liquids don't compress easily, so the ensuing hydraulic self-pressure can often be enough to burst the tank, or any closed-off feed plumbing downstream of the tank.

To prevent such an accidental hydraulic overpressure, then just as in nature's design of the egg, a small percentage of the tank volume is deliberately left free of liquid to allow for expansion with temperature. This gas pocket can then compress to absorb reasonable volume changes without over-pressuring the tank. For Nitrous, this 'head space' of gas is often created by situating a vent-pipe a little way below the top of the tank, so that the liquid never fills above the level of the vent (see chapter 3).

Large run-tanks benefit from either a pressure relief valve, which is a commercial safety valve designed to open at a set over-pressure, or a commercial burst disc (see page 95) that is designed to burst at a set over-pressure.

As every excess gram of run-tank matters, what is a reasonable minimum volume of head-space? Using tables of Nitrous properties with temperature (chapter 2) the Aspirespace run-tanks are designed to absorb the liquid expansion caused by a 10 degree C increase in Nitrous temperature after the vent is closed at the nominally British climate's 15 degrees C. This requires a 12% ullage (12% of the tank volume is vapour) which reduces to 0.7% at 25 degrees C.

The engine plumbing

Rigid pipework and welded joints are preferred for integrity, but swaged joints are better for modification or replacement of parts as they can be re-used: cutting a pipe and welding it again can introduce metal particles and other contamination into the pipework.

Threaded joints should be limited in number with oxidiser system plumbing, as the threads are difficult to keep clean. Where threaded joints can't be avoided, use PTFE tape to seal the thread, *not* sealing compound. Be careful not to let strands of PTFE come loose and fall into the pipework as they can block injector orifices or jam valves.

Plumbing should be properly clamped and supported to minimise flexing, chafing, abrasion, resonance, and mechanical strain that could lead to the pipework breaking. The supports should allow thermal expansion or contraction where necessary. Rigid pipes should be supported as close as possible to each bend in the pipe.

Connectors and fittings that are to be disconnected during normal operation should be provided with tethered end-plates, caps, plugs, or covers to protect the plumbing from contamination or damage when not in use. Keep it clean!

Avoid rust/corrosion coming into contact with Nitrous vapour as it can act as a catalyst for the Nitrous. This is particularly a problem with cheap zinc-passivated hydraulic fittings as their plating comes off or goes 'chalky' and the sulphide chemicals contaminating industrial-grade Nitrous can rot them over time. Stick to stainless steel and aluminium fittings where possible.

[A **catalyst** is a substance that can increase the rate of a chemical reaction, in this case the rapid decomposition of Nitrous (see page 102).]

Hybrid engines vibrate: use mechanical devices such as safety wire wrapping or bent metal tabs to prevent nuts unscrewing themselves.

Be sure to proof-test and leak test the plumbing.

The combustion chamber

As well as having suitable internal thermal insulation (such as Tufnol phenolic composite tube) the combustion chamber must not leak otherwise a plume of hot, escaping gas will erode the leak far wider and cause a burn-through. So leak test the chamber by plugging the nozzle. Viton rubber O-rings are sufficient to prevent leaks in Nitrous engines, but they melt at lowish temperature so should be shielded from the hot chamber gas by a covering of insulation (this traps a pocket of cool air which protects the O-ring).

Of course, the chamber is also a pressure vessel so needs to be hydro-tested as was done with the tanks (see earlier).

Nitrous leaks

Leaks show up in any pipe joints carrying the liquid phase of Nitrous as regions covered in ice; the Nitrous sucks heat out of the atmosphere as it leaks out to atmospheric pressure and vaporizes, freezing the water-vapour in the air around the leak. It'll freeze your hands or face too if they're near a leak: *wear goggles and gloves when you work with Nitrous.*

Terminal flatulence

We want to use the highest hybrid combustion chamber pressure possible to get the largest specific impulse, as close to the run-tank pressure as possible, but there's a practical limit. A common trend is to use too large a hole/holes for the injector orifices in order to minimise the pressure-drop across the injector, but this is a very dangerous practice.

The pressure drop is there partly to prevent very hot combustion chamber gasses having the potential to flow back upstream into the feed system or tank if there is a chamber pressure pulse. Too low a pressure drop across the injector encourages such audible forward-reverse flow oscillations in the chamber: *screaming hybrids are dangerous*. Hot gasses could get back into the feed system and start a fire (or decompose the Nitrous there).

As the Nitrous flows from the run-tank feed pipe into the injector orifices, the cross-sectional area available to the flow drops sharply. The mass continuity equation says that as the density doesn't change much, the flow velocity gets greatly increased. This increase in flow kinetic energy robs pressure energy from the flow: the pressure drops sharply (see appendix 3, Bernoulli's equations).

Usefully, this sudden drop in pressure (ΔP) causes the majority of the Nitrous to flash into vapour, so what comes out of the injector is a high-speed froth. This means that less energy of combustion has to be wasted in vaporising all of the Nitrous into gas so that it will burn with the fuel.

As an analogy, you can think of the injector as a weir on a river. The sudden drop in the height of the water as it flows over the weir is like the sudden pressure drop at the injector.

You have to ensure that the weir is high enough so that any large 'water waves' hitting it from downstream can't flow backwards over the weir. So in practice you have to make the pressure drop ΔP between tank and chamber - which occurs at the injector - high enough that any *pressure* waves can't 'get back over the weir'.

'Humble' (see references) advises a pressure drop across the injector of 20 percent of the combustion chamber pressure at the end of the liquid phase of the burn when the tank pressure is at its lowest.

Make sure that the upstream feed pipe (between run-tank and injector) is suitably wide enough (has a large enough cross-sectional area) that the flow velocity within it is less than 10 metres/second (use the mass continuity equation from chapter 1) to minimise pressure losses in this pipe, otherwise you'll get an unwanted pressure drop along the pipe which reduces the injector pressure drop.

An inadequate injector pressure drop also causes higher combustion instability that leads to higher engine vibrations. SpaceShipTwo hybrid engine vibration levels were so bad during early phases of flight testing that they would have torn the engine apart towards the end of the burn. This was part of the decision to switch to nylon fuel.

Nitrous oxide decomposition hazard

Some years ago, a fatal accident occurred at the test site of a very large hybrid. A large flight-weight run-tank of Nitrous exploded. So what exactly happened?

The facts are that a large composite tank had been filled with Nitrous in preparation for a flow test of the hybrid injector. This test was cold: there was to be no ignition of the engine. However, a device was added downstream of the injector that was electrically powered.

During the flow test, this electrical device malfunctioned and reacted with the Nitrous flowing through it and exploded. This fired the injector rearwards into the composite tank which shattered. With the loss of the injector, a flame front would have travelled back into the tank.

The majority of the Nitrous spontaneously decomposed into nitrogen and oxygen gas, with a large release of heat from the decomposition (the casualties suffered severe burns) that raised the run-tank pressure. This over-pressurised the run-tank, and as there were no pressure relief devices on the tank such as a burst disc or pressure relief valve, the tank burst.

Although a lot of damage then occurred to the test area, it was significantly less damage than would have occurred had there been a detonation of the Nitrous. The definition of a detonation is that the flame front spreads through the Nitrous at supersonic speed. In fact no evidence has yet been uncovered to suggest that Nitrous will detonate.

The data suggests that spontaneous decomposition can occur, spreading through the Nitrous at a rapid pace, but significantly slower than a typical fuel/air deflagration.

It appears that only the Nitrous vapour is the culprit, liquid Nitrous doesn't appear to support continued decomposition because the surrounding liquid soaks the heat out of the reaction (this is called **quenching**): all attempts in the literature to ignite liquid Nitrous have failed.

It's probable that the damage was caused by a **BLEVE** event. This stands for Boiling Liquid Expanding Vapour Explosion.

When the Nitrous vapour decomposed and, along with the injector, ruptured the tank, the tank pressure suddenly dropped to the pressure of the atmosphere outside the tank. This caused all of the liquid Nitrous to flash-boil into vapour, expanding enormously in the process.

This expansion provided the energy for propagation of further cracks in the run-tank wall (shrapnelling) and then propulsion of these fragments of tank at very high velocity.

The expansion was further enhanced by a greater number of moles in the decomposed gasses and their increased compressibility factor (a so-called 'real' gas effect).

As this was happening, the now greatly increased amount of Nitrous vapour decomposed, causing a release of heat which caused further gas expansion.

Complete decomposition of the Nitrous in a run-tank will theoretically cause a twenty times increase in pressure, though the tank will have burst long before reaching that pressure.

[Note: Pressure relief valves are sized to release pressure fast enough to prevent the pressure from increasing beyond the strength of the vessel, but, for Nitrous, not so fast as to be the cause of an explosion. An appropriately sized relief valve will allow the liquid inside to boil slowly, maintaining a constant pressure in the vessel until all the liquid has boiled and the vessel empties. Basically, too large a relief valve orifice or burst diaphragm could cause a BLEVE event.]

Alternatively, there are suggestions that the run-tank was filled hours earlier then left to the mercy of the Sun: the test was then run at the hottest part of the day. The ambient temperature was reported as 40+ degrees C.

Nitrous goes supercritical at 36 degrees C (chapter 2) so the tank or plumbing could've contained supercritical Nitrous. Supercritical Nitrous is very susceptible to pressure-shock which will result in a very high velocity decomposition during which temperatures can exceed 3,000 degrees C. This may also have caused the demise of the MARS team's Deimos Odyssey Nitrous hybrid in the heat of the Blackrock desert USA.

It turns out that Nitrous is a good solvent, particularly if it's supercritical. The run-tank used a composite tank liner, so bare composite resin was in contact with the Nitrous, as well as the waxy deposits that you get on the surface of composites. Nitrous can dissolve the hydrocarbon binders in the resin. This became the fuel, or possibly the catalyst (see page 101) for the decomposition. Plastic saturated with Nitrous can decompose 'explosively' when ignition energy is provided.

Static discharge

What provided the ignition energy?

It was a hot, very dry day in the test area that day, which is conducive to the build-up of static electricity. Our Nitrous hybrids typically have metal tanks and combustion chambers, so they're earthed when in contact with our metal test-stands. But this hybrid had a composite tank and chamber. Assuming that the injector plate was metal, then this could have built up a large static charge as the Nitrous flowed through it.

The electrical conductivity of Nitrous is low enough that with flowing Nitrous it is theoretically possible to produce a large enough static discharge to initiate decomposition.

As liquid Nitrous flows through an injector, Nitrous vapour appears. This vapour can be easily ignited by an electrical spark at typical tank pressures, and even a very feeble spark will do it (0.14 Joule). One rocketeer from the Arocket rocketry forum recounts: "I can tell you from personal experience that a big spark will set off Nitrous."

Electrical arc

The other candidate for ignition was the device attached downstream of the injector. It appears to have exploded, perhaps oil or grease plus electricity reacted with supercritical Nitrous.

Scale effect

Another issue with the hybrid was its very large size. The radius of the feed pipes of many small amateur hybrids are smaller than the **quenching distance**. This is a size (around 7 millimetres for pressurised Nitrous, though it changes with pressure) below which the metal walls of the pipe are near enough to soak up any rogue heat energy from decomposition, stopping the reaction dead. Researchers could only get Nitrous vapour to sustain a reaction in a ¼ inch radius (6.35 millimetre radius) metal pipe by heating it prior to ignition (204 degrees C and 55 Bar). In a 1/2 inch radius pipe and larger, there was no quenching at typical Nitrous pressures.

Contamination

The presence of even a small amount of fuel or catalyst (see page 101) material in tanks and feed plumbing can greatly reduce the energy threshold required for initiation of Nitrous decomposition. A mixture of Nitrous and 9% hydrocarbon (ethanol) initially at only 40 degrees C, exploded in a lab.

The way to deal with contamination issues is to give the system a damn good clean; equipment must be thoroughly degreased before use. See 'degreasing procedures' below.

Adiabatic self-compression

This is also called 'water hammer' (even though there's no water), and until recently we all thought it only occurred with pure high-pressure oxygen (Gox).

If a run-valve is opened suddenly, Gox rushes down the feed pipe until it hits another shut valve or obstruction such as the injector. The Gox's momentum piles it up against the shut valve, and the temperature rises purely because of the compression. This self-heating can reach the ignition temperature of the pipework, and start a fire. In Nitrous' case, the self-heating can reach its auto-decomposition temperature.

The way to prevent adiabatic compression is to avoid sudden rushes of oxidiser. Open run-valves slowly and reduce the dead-space downstream of valves: dead volumes in the feed lines (e.g. a tee fitting) are prone to adiabatic compression. Reduce the pressurisation rate of your run-tank as you fill it to no more than 20 psi per second to avoid adiabatic compression. The way to do this is to reduce the diameter of both your fill pipe and your run-tank vent.

Reverse flow: flashback through Nitrous vapour

Flashback of hot chamber gasses into the Nitrous vapour *after* all the liquid has run out is thought to be a greater hazard than problems before the liquid runs out. This is because there's much more vapour in the feed system, and this vapour goes all the way up and into the run-tank, so the tank can overpressure too. Hybrids have exploded when the liquid Nitrous has run out, and a chamber pressure spike has reversed the flow.

Inadvertent liquid engine, a hard start

One of the biggest bangs of recent years in U.K. rocketry occurred when a Nitrous hybrid engine accidently became a liquid engine, and then blew up. The Nitrous *must not* be in liquid form once it's inside the combustion chamber because of the danger of it pooling in corners of the engine; molten plastic fuel may be pooling in the same corner.

In this case, the Nitrous was injected tangentially into the chamber instead of axially for several promising technical reasons, but centrifugal effects re-compressed the Nitrous, and at the higher pressure it reverted into a liquid again.

A design oversight allowed molten plastic to occupy the same area, and it did the classic liquid rocket engine 'hard start' upon ignition. This bang was compounded by the fact that the feeble igniter was situated at the nozzle instead of where you ought to put it: close to the injector, so the chamber could happily fill with unburnt liquid Nitrous.

Decomposition

It's instructive to review the decomposition process. Nitrous decomposition is a marginal reaction with just enough heat released to sustain itself. It is easily quenched: either by nearby metal pipe walls or a dilutent gas added to the vapour acting to absorb heat.

Dilutent

Small concentrations of dilutent gas added to the Nitrous vapour increase the ignition energy of the vapour, making the mixture extremely difficult to ignite at dilution levels greater than 30%. Adding extra gas to the run-tank is sometimes called 'supercharging', and has the side effects of raising the tank pressure slightly, and reducing the tendency for two-phase flow in the feed system.

Dilution with nitrogen is expected to have a dilutent effect, and helium is expected to be better. It should have the same effect at lower dilutions because heat gets conducted away more easily in helium as it has a higher thermal conductivity. Research suggests you need four times less helium compared to nitrogen.

Flame traps

There's been talk on the amateur rocketry web forum Arocket of the possibility of adding a flame trap to the feed line just upstream of the injector. (Technically it's actually a quenching trap.) Nitrous has a much larger quenching distance than typical fuel/air mixtures, so a grid of stainless steel tubes put streamwise into the feed line could be able to stop any decomposition that starts at the injector. The quenching distance for Nitrous is around 7 millimetres, so I reckon that a grid of 5 millimetre diameter tubes will work, and the tubes are large enough not to cause much of a pressure drop. But I must stress that I haven't found any evidence of anyone having tried a flame trap with Nitrous. This is a (potentially hazardous) experiment that needs doing.

Degreasing procedures

With oxidiser systems (Nitrous, Lox, etc.) it is vital to remove all traces of dirt and grease upstream of the combustion chamber when assembling the plumbing to prevent fire or Nitrous decomposition.

Clean parts in Isopropyl alcohol (in an ultrasonic tank if possible) then blow out the remaining Isopropyl alcohol with clean compressed air (<u>not</u> from an oil-lubricated air-compressor) or nitrogen, or wipe off with clean lint-free tissues/wipes.

For parts with inaccessible internal areas, put the part under vacuum or in a warm oven to ensure all the Isopropyl alcohol evaporates.

You can use Amberclene FE10 or aqueous alkali-based degreasers too. Or one can use dilute (20%) hydrogen peroxide to remove traces of grease from oxidiser tanks where it isn't possible to physically wipe the surface. <u>This should only be done once any bulk contamination has been dealt with or it'll ignite!</u>

If your budget won't permit the above, hybrid parts can be degreased using domestic cleaning products (detergents) in water, automotive brake-part cleaner, or preferably Isopropyl alcohol, then dried in a warm oven for a few hours.

If degreasing elastomeric seals (e.g. O-rings, washers) then check material compatibility first and don't leave them soaking in solvents for long periods as they can swell-up/soak up the solvent. Instead, rinse for a few seconds then wipe dry.

The use of Isopropyl alcohol (and other organic solvents) has safety implications: it has a flammable vapour which is heavier than air but the vapour pressure is quite low so explosion risk is rarely an issue, but don't use it near sources of ignition (including electrical contacts like switches, thermostats, or relays.) Don't inhale the vapour, and use in a well ventilated area. Also, it is mildly toxic when absorbed through the skin, so wear plastic gloves.

Nitrous material compatibility

As a general guide, materials that are oxygen-compatible are suitable for Nitrous.

It's worth pointing out that I haven't seen any evidence for a catalyst (see page 101) for Nitrous that works at room temperature, they all seem to need elevated temperatures.

Metals

- Aluminium, Stainless Steel: satisfactory.

- Copper and its oxides (and brass/bronze), nickel, and platinum, are highly catalytic with (promote decomposition of) Nitrous especially at elevated temperatures, so do not use these.

- Carbon steels are corrosive in presence of moisture: do not use corrosion-prone metals because, for example, iron oxide (rust) is a catalyst. Avoid rust contamination from steel fill-tanks; use stainless steel filters in the fill line to catch any rust particles.

Plastics

- Fluoroplastics such as Polytetrafluoroethylene (PTFE), Polychlorotrifluoroethylene (PCTFE), Fluorinated ethylene propylene (FEP), Perfluoroalkoxy polymer resin (PFA), are satisfactory.

- Other plastics are an ignition hazard.

- Nitrous can saturate plastics and composites turning them into 'explosives': HTPB fuel saturated with Nitrous can be 'explosive'.

Elastomers

- Think carefully about what material to use as your run-tank O-rings. Just because commercial HPR hybrids use 'rubber' O-rings doesn't automatically make it a good idea.

- Fluorocarbon-coated O-rings: use these where possible. Unfortunately, they don't stretch very much.

- Buna-N and neoprene degrade in Nitrous liquid after several days.

- Butyl (isobutene - isoprene) rubber (IIR), Nitrile rubber (NBR), Chloroprene (CR), Ethylene - Propylene (EPDM): these are a possible ignition hazard and experience significant swelling when immersed in Nitrous.

- Silicon (Q): is satisfactory.

- Silicone saturated with Nitrous is an impact-sensitive explosive.

<u>Lubricants</u>

- Nitrous can form an explosive with many hydrocarbons and lubricants.

- Fluorocarbon based lubricant (krytox): is satisfactory.

- Hydrocarbon based lubricant: is *definitely not recommended*, it's a possible ignition source.

Recommendations

- Minimise the number of persons present in the test area whenever Nitrous is being loaded/is loaded into the run-tank or whenever Nitrous is flowing.

- Avoid decanting Nitrous in small, confined areas as Nitrous is an anaesthetic. If using Nitrous in a confined space use an oxygen sensor to monitor dangerous concentrations of Nitrous which could be inhaled.

- Use a large-volume (large diameter) burst disc on the run-tank. This is mandatory for manned systems. But not so large as to cause a BLEVE (see page 103).

- For engine static testing, keep the run-tank upright so only liquid gets to the injector at ignition.

- Keep the plumbing *very* clean: degrease the plumbing as even very small traces of contamination can cause problems.

- Avoid the use of catalytic materials.

- Nitrous is a good solvent of hydrocarbons: such as grubby fingermarks, O-ring polymers, and valve seals.

- Venting the Nitrous through the combustion chamber should be avoided at all costs.

- Dilution of the Nitrous vapour in the run-tank is recommended.

- Avoid hard starts: fire the igniter before admitting the Nitrous into the combustion chamber.

- Properly earth all ground support equipment (fill-tanks and fill lines), and keep mobile (cell) phones and laptop Wi-fi switched off when anywhere near Nitrous. Persons handling Nitrous should preferably be earthed.

- Construct tanks and lines out of conductive material. For composites, work out a way of dissipating static electricity (aircraft composites have metal foil as their outer layer to dissipate lightning strikes. Another approach is to embed carbon nanofibres or nanotubes within the resin.)

- No part of the engine should have a resistance of more than one ohm between it and its adjacent parts so that potential differences (voltages) cannot build up.

- Avoid adiabatic self-compression: minimise feed-line dead volume downstream of the run-valve, and slow the valve opening.

- Prevent back-flow of igniter gas through the injector.

- Reduce the pressurisation rate of your run-tank as you fill it to no more than 20 psi per second.

- Use small enough injector orifices and widen the nozzle throat to get a good pressure drop across the injector right through the burn (a drop of 20% of the combustion chamber pressure as the liquid phase runs out).

- Develop safety procedures, and pre-chill the pump, if you pump Nitrous from the fill-tank to the run-tank.

- Use pressure relief valves/burst discs on all trapped volumes of liquid Nitrous to prevent hydraulic overpressure.

- Review all moving parts in the Nitrous system (regulators, valves) for friction, impact, and static discharge.

- Avoid eddies or stagnation zones in the feed pipework due to sudden changes in cross-section. These can act as flame-holders which prevent any decomposition from being flushed safely downstream.

Chapter 9: Design the engine

You've learnt a lot about rocket engines, so it's now time to design our large hybrid engine. First, we'll find out how large the spacecraft tank and hybrid combustion chamber have to be (i.e. how much propellant is required to reach just over 100 kilometres apogee). We'll find these by using a trajectory simulator.

Once the hybrid is built (in chapter 10), you then have to test-fire it (in chapter 11) to see if it works as you expected.

How big?

The first question is how large must our hybrid be? How much thrust and how much propellant will we need to reach just over 100 kilometres altitude if we start from a gas balloon at 21.3 kilometres (= 70,000 feet)?

First we need to work out how much mass to carry up there.

Based on my experience of aircraft design, I'm estimating that a one-seater spacecraft will have an empty mass (termed the **zero-fuel mass**) of about 400 kilograms. This is the spacecraft's mass before filling it with propellants. This zero-fuel mass includes your body's own mass (you're the payload), and the mass of the empty propellant tank and empty hybrid engine.

Before we can start building our hybrid (in the next chapter) we first have to write a trajectory simulation to describe the trajectory. After running the sim several times, we can home-in on the required propellant mass to reach 100 kilometres.

Initial estimate

It's useful, before you write a sim, to have a feel for the numbers you're going to put into it, and the likely answers the sim will give you. This gives you confidence that the sim has been coded correctly so is not giving you wildly inaccurate answers.

We'll use an energy equation to work out how fast we need to get the spacecraft ascending for it to coast up to 100 kilometres after engine burnout. Then we'll use the rocket equation (see below) to work out roughly how much propellant we'll need to burn to get up to this speed.

We need to guesstimate the altitude at which the engine burns out. I have enough experience of similar suborbital trajectories to tell you that a realistic burnout altitude is around halfway between launch altitude and apogee, i.e. halfway between 21.3 kilometres and 100 kilometres, which is just over 60 kilometres. Then coast from this altitude up to 100 kilometres.

We'll work out the gravitational potential energy of our spacecraft at 100 kilometres, then we'll let it fall from an initial zero speed at 100 kilometres down to 60 kilometres and so work out how much of the original potential energy got converted to kinetic energy during the fall. This will give us the speed the spacecraft will be doing as it falls past 60 kilometres.

Reversing this process: the speed the spacecraft has as it falls past 60 kilometres is exactly equal to the speed that we'd have to throw the spacecraft *upwards* at 60 kilometres to just reach 100 kilometres.

So over the fall from 100 kilometres to 60 kilometres, the loss of specific gravitational potential energy of the spacecraft is (see chapter 1):

$$\Delta e_{potential} = \frac{\mu}{R_E+60} - \frac{\mu}{R_E+100} = 382.285 \; kiloJoules \qquad \text{equ. 9.1}$$

where μ = 398,600,500 km³s² is the Earth's gravitational constant, R_E is the mean radius of the Earth (6,378.14 kilometres), and h is the height you're at above the Earth's surface *in kilometres*.

This loss of specific potential energy is equal to the gain in specific kinetic energy of the spacecraft as it falls past 60 kilometres.

Recall that specific kinetic energy $= \frac{1}{2}V^2$ so:

$$\frac{1}{2}V^2 = 382,285 \, Joules \quad \text{equ. 9. 2}$$

Rearranging:

$$V = \sqrt{2 \times 382,285} = 874 \text{ metres per second} \quad \text{equ. 9. 3}$$

So this is the speed that we have to throw the spacecraft *upwards* from 60 kilometres in order to just reach 100 kilometres apogee.

[Note that we've just performed the calculation for just one kilo, but the answer is the same for however many kilos comprise your spacecraft.]

The burn will take about a minute so that the acceleration on your body isn't too high. During this long burn, gravity will have all this time to pull down on the vehicle to create a large Gravity loss (see equation 1.35).

The Gravity loss for a vertically ascending vehicle is equal to g times the burn time. We have to choose a representative gravity value for the burn, which we'll calculate as that occurring at 40 kilometres, which is roughly halfway through the burn (roughly halfway between 21.3 kilometres and 60 kilometres).

From the gravity equation (chapter 1):

$$g = \frac{\mu}{(R_E + h)^2} \quad \text{equ. 9. 4 in metres/second}^2$$

Where again $\mu = 398,600,500 \text{ km}^3\text{s}^2$ is the Earth's gravitational constant, R_E is the mean radius of the Earth (6,378.14 kilometres), and h is 40 kilometres.

So:

$$g = \frac{398,600,500}{(6378.14+40)^2} = 9.68 \text{ metres/second}^2 \quad \text{equ. 9. 5}$$

This Gravity loss is actually an extra velocity that the rocket will have to make up.

We don't know the burn time of the engine. As a first guess, we'll say it lasts 50 seconds. So the Gravity loss is:

$$\Delta V_{gravity\,loss} = g \, burn_{time} = 9.68 \times 50 = 484 \text{ metres per second} \quad \text{equ. 9. 6}$$

As well as the Gravity loss there will of course be a Drag loss (see equation 1.36) which is another velocity the engine will have to make up. We've no idea how large this Drag loss will be, however for a good trajectory with low Drag and Gravity losses, the Drag loss is usually about equal to the Gravity loss. We'll assume it is.

So:

$$\Delta V_{drag\,loss} = 484 \text{ metres per second} \quad \text{equ. 9. 7}$$

Therefore the rocket has to make a total change in velocity of:

$$\Delta V = 874 + 484 + 484 = 1,842 \text{ metres per second} \quad \text{equ. 9. 8}$$

The rocket equation

Now we need to work out how much propellant we'll need to burn to reach this velocity. This calculation, which is called **the rocket equation**, was conceived by one of the world's first rocket scientists, Kontanstin Tsiolkovsky. (Pronounced 'Zee-ol-covskie'.)

His equation is:

$$\Delta V = C_e \ln\left(\frac{spacecraft\ mass\ at\ ignition}{spacecraft\ mass\ at\ burnout}\right) \quad \text{equ. 9.9}$$

where the fraction:

$$\frac{spacecraft\ mass\ at\ ignition}{spacecraft\ mass\ at\ burnout} \quad \text{equ. 9.10}$$

is known as the **Mass ratio**.

[Note that Sutton's famous rocket book is all alone in defining the Mass ratio the other way up, i.e.

$$\frac{spacecraft\ mass\ at\ burnout}{spacecraft\ mass\ at\ ignition}$$ only George Sutton knows why!]

[A British researcher William Moore came up with the rocket equation first, but at the time nobody saw his work which is what happens if you work for the military: secrecy.]

ΔV is the increase in the vehicle's velocity after all the propellant is burnt and, as before, C_e is the engine's effective exhaust velocity (chapter 5), *ln* is the natural logarithm function on a scientific calculator (or log-e).

The spacecraft mass at ignition is equal to the vehicle's tanks-empty mass (which occurs at burnout) plus the mass of the propellants.

We need to turn the rocket equation around. We've already worked out what the ΔV needs to be to get up to 100 kilometres. What we want to know is what the propellant mass has to be.

From above, the rocket equation is:

$$\Delta V = C_e \ln(Mass\ ratio) \quad \text{equ. 9.11}$$

First, divide both sides by C_e:

$$\frac{\Delta V}{C_e} = \ln(Mass\ ratio) \quad \text{equ. 9.12}$$

Then use the exponential function to reverse the logarithm:

$$Mass\ ratio = exp\left(\frac{\Delta V}{C_e}\right) = exp\left(\frac{1,842}{2,682}\right) = 1.987 \quad \text{equ. 9.13}$$

So we need a Mass ratio of roughly two. (Recall that we worked out on page 86 that our effective exhaust velocity was 2,682 metres per second.) This is a reasonable first guess, as most similar air-launched suborbital missions require a Mass ratio of around two.

Now the **Mass ratio** = $\dfrac{spacecraft\ mass\ at\ ignition}{spacecraft\ mass\ at\ burnout}$ = $\dfrac{empty\ mass + propellant\ mass}{empty\ mass}$

$$= \dfrac{empty\ mass}{empty\ mass} + \dfrac{propellant\ mass}{empty\ mass} = 1 + \dfrac{propellant\ mass}{empty\ mass} \quad \text{equ. 9. 14}$$

Then subtracting 1 from both sides:

$$\boldsymbol{Mass\ ratio - 1} = \dfrac{propellant\ mass}{empty\ mass} \quad \text{equ. 9. 15}$$

And finally, multiply both sides by *empty mass*:

$$\boldsymbol{propellant\ mass = empty\ mass\ (Mass\ ratio - 1)} \ = 400 \times (1.987 - 1) \ = 395\ kg \quad \text{equ. 9. 16}$$

So very roughly, we need about 400 kg of propellants.

We could make a better estimate of the Gravity loss by calculating the burn time correctly as in fact we do later, but we would still only be guessing the Drag loss. And we're not taking account of the beneficial higher effective exhaust velocity towards the end of the burn where the ambient air pressure is lower. It's better to move on and program a trajectory simulation, which is what we'll now do.

Numerical integration

Our first trajectory sim is simple enough to write on a computer spreadsheet such as Microsoft Office's Excel. We'll do this shortly, once we've worked out what we need to put into the sim.

Starting at the top of the spreadsheet, we'll increase time by half a second between each row and the next, and we'll keep adding rows (increasing time) until we reach apogee. There is a good reason for stepping the time forward in little chunks (Δ time) and it's called **numerical integration**.

Numerical integration is integration (chapter 1) performed using a computer.

Our trajectory sim will be used to work out the final effect - after the whole of the burn time is passed - on our spacecraft's velocity. This velocity is caused by the acceleration from our rocket's thrust: we want to know the spacecraft's velocity at engine burnout.

This is simple to calculate if the acceleration *doesn't* change over time (is constant with time). If it *is* constant, then the change of velocity of the vehicle (Delta V, 'Δ V') after some time interval Δt has elapsed is simply:

$\Delta V = a\ \Delta t$ equ. 9. 17

where 'a' is the constant acceleration. This is just simple multiplication.

Unfortunately, our hybrid engine doesn't produce a thrust that is constant with time, so the acceleration that it gives isn't constant with time. That's unfortunate as we'd still prefer to use the simple ΔV equation.

However, the acceleration only changes *slowly* as time passes.

We divide the burn time into numerous little slices of time (small Δt) called **time-steps**.

We can then say that the acceleration changes only slightly over the short time of a time-step, so we can approximate that the acceleration is actually *constant* over each small time-step.

So we assume that the graph of the acceleration (the curved top line here) is actually a series of flat steps formed by the roof of each rectangle as shown here: Each flat roof is caused by a constant acceleration.

So over each time-step Δt, we can indeed use the simple ΔV equation:

$$\Delta V = a\,\Delta t$$

where 'a' is now a constant approximation for the slowly changing acceleration that occurs at that particular time. (For the next time-step, 'a' will be a new constant value.)

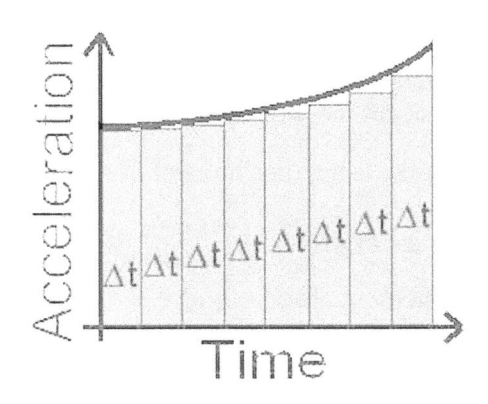

An $a\,\Delta t$ gives us a rectangle with area $a \times \Delta t$.

This only gives us the change in velocity over a small time-step, but we can then simply add together all the individual ΔV's occurring over all of the little Δt's and this will give us the total change in velocity occurring over the time of all the Δt's added together.

Okay, this is rather a fudge, we'll only get an approximation to the correct answer for the total ΔV at burnout, but with small enough steps, the sim answer is surprisingly accurate.

So numerical integration is simply the adding together of the areas of all the rectangles in the above graph. With modern computers, especially the new 64-bit machines, we can use exceedingly narrow rectangles of time, time-steps only tiny fractions of a second wide, to get extremely close to the right answer.

But it's not always worth it: the narrower the rectangles are, the more of them you have to add together, and this takes computer time. What you want is a time-step that is narrow enough to give you an *accurate enough* answer, but doesn't take hours to run the sim.

A way to judge what value makes a suitable Δt is to halve it and then re-run the sim with this smaller time-step: does the final answer change noticeably? If it does, halve it again and so on until the change is too small to bother about (after all, we only need to know our apogee to an accuracy of about 500 metres: if our sim's apogee answer is 500 metres out then it really doesn't matter.)

For the trajectory sim we shall put together shortly, a Δt of 1/2 a second width gives a good enough answer.

We used rectangles in the above graph to keep the maths simple, but you can imagine other shapes that would work better. For example, you could add little triangles to the top of each rectangle to fill in the gap between the roof of the rectangle and the top curve.

There are indeed other integration formulae that calculate the area more precisely and so give a more accurate answer for whatever the time-step size you're using, but they can be much more complex to program. Having said that, the more complex integration formulae are much less prone to 'exploding' (the results of the integration become erroneously enormous) so the extra programming effort is often worth it.

In chapter 1 we saw that 'velocity is the time integral of acceleration':

$$\Delta V = \int_{start\ time}^{end\ time} a\,dt \quad \text{equ. 9.18}$$

So if we numerically integrate acceleration, we get the total increase in velocity, but only the change in velocity that has occurred up to the end time-point (the point on the horizontal time axis) that we cease to add more rectangles.

In chapter 1 I suggested burnout as a handy time to stop integrating at but in this sim, we'll actually just keep on integrating. We'll continue to add more 1/2-second time-step widths to the total time in every row of the sim spreadsheet so that time increases by 1/2 second between one row of the spreadsheet and the next. Once the spacecraft has coasted all the way up to its apogee, then we can stop integrating and the sim is finished.

Velocity is the time integral of acceleration, but similarly, distance (displacement) is the time integral of velocity:

$$\Delta height = \int_{start\ time}^{end\ time} V\ dt \quad \text{equ. 9.19}$$

Furthermore, **total impulse** is the time integral of thrust:

$$\textbf{Total impulse} = \int_{start\ time}^{end\ time} F\ dt \quad \text{equ. 9.20} \quad \text{where } F \text{ is the thrust.}$$

[Total impulse gives a good measure of how much ΔV a rocket will create by burnout; you can think of it as 'the total effectiveness - over the time of the burn - of the varying thrust'. Total impulse is numerically equal to the area 'under' the thrust curve.]

And if we look at the area under the graph of nozzle mass flow rate $\dot{m}_{nozzle}$ (see page 115), we get the total propellant mass just before ignition, so total propellant mass is the time integral of nozzle mass flow rate:

$$\textbf{Total propellant mass} = \int_{start\ time}^{end\ time} \dot{m}_{nozzle}\ dt \quad \text{equ. 9.21}$$

The hybrid rocket engine

Within this trajectory sim we need to simulate the hybrid engine. This can end up being surprisingly complex, so the first task is to create a simplified version of a hybrid's **thrust curve**, which is the rocketry term for the graph of thrust versus time. Even with such a simple approximation of the engine, the sim gives us a good guess at the required size of our hybrid.

The mass flow rate curve and the thrust curve

Our hybrid is powered by a tank of Nitrous oxide. When Nitrous empties itself from a tank, the vapour phase expands to fill the increasingly empty space at the head of the tank, and as it expands, it chills itself. This cooling drops the tank pressure (the vapour pressure, see chapter 2) continuously throughout the burn until at burnout the Nitrous is down to about 2/3rds the vapour pressure it started at.

Interestingly, it always drops down to about 2/3rds no matter how large or small the run-tank is, or how long the burn is, or what the vapour pressure was to start with. (This does suppose that the run-tank was almost full of liquid Nitrous to start with.)

We'll assume that the Nitrous starts out at 20 degrees C which gives a start tank pressure of 51 Bar (see the end of appendix 2).

[Note that as the temperature of the atmosphere at 21.3 kilometres is considerably below 20 degrees C then the run-tank will require an electrical heating blanket.]

The continuing drop in tank pressure during the burn means that the force forcing the Nitrous out of the tank and into the chamber is dropping as time goes by. So the mass flow rate of Nitrous into the combustion chamber steadily drops throughout the burn. So the balance between what flows into the chamber ($\dot{m}_{in}$) and what flows out ($\dot{m}_{out}$) is upset as less is flowing in.

So from equation 4.17 the combustion chamber pressure therefore steadily drops throughout the burn, and so the nozzle mass flow rate decreases too.

With our Aspirespace Nitrous hybrids, the nozzle mass flow rate at burnout has dropped to about 70% of the value it started at.

What about the exhaust velocity?

We'll assume that the exhaust velocity stays the same throughout the burn for this very high altitude flight.

Is this a reasonable assumption?

The back-pressure of the atmosphere on the nozzle is getting progressively lower with time as the rocket gains altitude, but at the same time the combustion chamber pressure is getting progressively lower, so we'll assume that the *difference* (ΔP) between these two pressures stays the same, which makes the exhaust velocity stay about the same.

With the mass flow rate dropping as the chamber pressure drops with time, and the effective exhaust velocity staying the same, then when you multiply them together to get the thrust, the thrust will drop continuously with time too.

The graph of combustion chamber pressure dropping with time is near-enough a straight line with Nitrous hybrids, so our simple thrust curve approximation will be a straight line.

So the nozzle mass flow rate graph and the thrust curve within our sim will look like this, a downward slope with time:

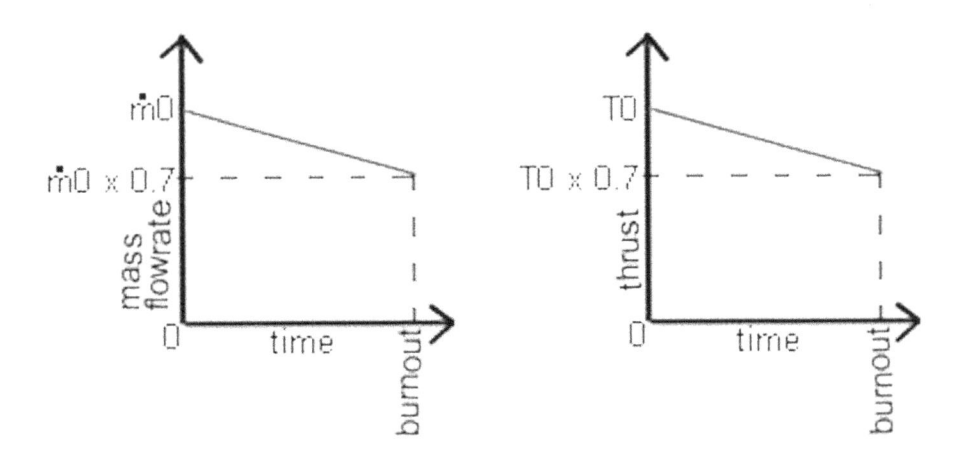

Because the exhaust velocity is constant with time this makes the two curves identical (remember, thrust = mass flow rate times exhaust velocity).

How much propellant mass?

At the start of our sim we need start values: start velocity is zero, start height is 21.3 kilometres, start acceleration is zero. But what is our start mass? We need a first guess of our start propellant mass.

I reckon that <u>353</u> kilograms of propellants will do the job of lifting our 400 kg zero fuel mass vehicle to just over 100 kilometres apogee.

This isn't a guess, what I've done is actually program the sim into a spreadsheet as we do over the rest of this chapter. I then obtained this mass by running the sim repeatedly with different masses (starting at our first-guess mass from earlier) until I got the correct apogee. That's how you use a sim in practice, so I'm just saving you time.

Note how that compares to our earlier estimate of 400 kilograms, the discrepancy is due to a lower Drag loss in reality from what we estimated.

Final vapour mass

Note that there is something peculiar about just-subcritical propellants emptying from a tank. Roughly 12% of the start propellant mass doesn't get burnt in the usual way.

For example, when all of the liquid Nitrous oxide has drained out of the tank, there's still about 12% of the start propellant mass left in the tank in the form of Nitrous vapour. This was originally liquid that the Nitrous self-vaporized in order to try and raise the vapour pressure back up. This vapour *will* empty itself into the combustion chamber and get burnt, but the mixture will be far off best mixture, so the thrust is reduced.

For the sim, we want to find a propellant mass X that when added to the final vapour mass '$mass_{vapour}$' gives us our required start propellant mass of 353 kilograms.

$$mass_{vapour} = 0.12(propellant\ mass) \qquad \text{equ. 9.22}$$

and:

$$mass_{vapour} + X = propellant\ mass \qquad \text{equ. 9.23}$$

Substituting for vapour mass:

$$0.12(propellant\ mass) + X = propellant\ mass \qquad \text{equ. 9.24}$$

gives:

$$X = propellant\ mass(1 - 0.12) = 353(1 - 0.12) = 310.6 \text{ kilograms} \qquad \text{equ. 9.25}$$

and:

$$mass_{vapour} = 0.12(propellant\ mass) = 0.12(353) = 42.4 \text{ kilograms} \qquad \text{equ. 9.26}$$

The Nitrous vapour burn

As the vapour empties from the tank after the liquid is gone, its pressure drops sharply in just a few seconds, so the combustion chamber pressure rapidly falls to low pressure.

Because the majority of the vapour-only burn occurs at low combustion chamber pressure, its specific impulse can be low *unless* the engine is at very high altitude by the time the vapour burn occurs, (which indeed happens with our hybrid) because a rocket thrusting into a near-vacuum *does not* lose efficiency due to low combustion chamber pressure.

From appendix 3, the exhaust velocity is:

$$V_e = \sqrt{\left(\frac{2R\gamma}{\gamma-1}\right) T_{chamber} \left[1 - \left(\frac{P_{exit}}{P_{chamber}}\right)^{\frac{\gamma-1}{\gamma}}\right]} \qquad \text{equ. 9.27}$$

If P_{exit} is almost zero, then $\frac{P_{exit}}{P_{chamber}}$ is also zero, and the $\left[1 - \left(\frac{P_{exit}}{P_{chamber}}\right)^{\frac{\gamma-1}{\gamma}}\right]$ term is at a maximum of 1.

In a previous incarnation of this book, I ignored the vapour burn for simplicity. This resulted in a very non-optimal hybrid that was wasteful of propellant. I apologise to readers of the former book because the required propellant mass for this mission was overestimated. In fact, the vapour burn is not negligible: at our launch altitude, it gives a Total impulse of roughly 13% of the complete burn Total impulse (liquid burn plus vapour burn).

This 13% varies with altitude, because at higher altitudes the vapour burn becomes relatively more efficient as we've just noted.

Altitude Kilometres	Vapour burn percentage of Total Impulse
0	11.5
5	11.9
10	12.3
20	12.6
30	12.8
Vacuum (area ratio of 100)	13.2

How much thrust?

Now we pick some start thrust. It has to be more than the weight caused by **353 + 400 = 753 kilograms** (which from chapter 1 is **753 × 9.81 = 7,387 Newtons**) or we won't go upwards.

Also, we know that propellant mass is continuously burnt and lost overboard through the nozzle, so the spacecraft will get lighter as time passes.

Recall Newton's 2nd law:

$$acceleration = \frac{force}{mass} = \frac{rocket's\ thrust}{mass} \quad or \quad a = \frac{T}{m} \quad \text{equ. 9.28}$$

Think about the maths of this equation here. Suppose the mass on the bottom of this fraction is getting smaller at a faster pace than the thrust on the top of the fraction is dropping. What happens is that the acceleration increases. So the acceleration on our spacecraft could steadily increase until burnout, and in fact it does (taking burnout to mean that time when all the liquid Nitrous has been used up).

So although the thrust is dropping with time, the acceleration will get larger with time. If we start with too high an acceleration, it could end up getting way higher than the 4 gees your body can comfortably take. But we don't want much less than 4 gees, because the lower the acceleration, the bigger the Gravity loss (chapter 1).

So we'll start with a thrust that is 3.3 times the weight, which is 24,200 Newtons. (You'll find out later why I picked this particular number.)

Remember from a few pages back that at the end of the burn, the thrust will be 0.7 times this value:

24,200 × 0.7 = 16,940 Newtons. We'll call this simulation equation number 1 (**Sim equ.1**)

How much nozzle mass flow rate?

Rearranging the thrust equation:

$$thrust = nozzle\ mass\ flow\ rate \times effective\ exhaust\ velocity$$

we get:

$$start\ nozzle\ mass\ flow\ rate = \frac{start\ thrust}{effective\ exhaust\ velocity} = \frac{24200}{2682} = 9\ kilograms/second \quad (\textbf{Sim equ.2})$$

Remember that at the end of the burn, the nozzle mass flow rate will be 0.7 times this value, which is:

9 × 0.7 = 6.3 kilograms per second   (Sim equ.3)

So the nozzle mass flow rate graph and the thrust curve look like this:

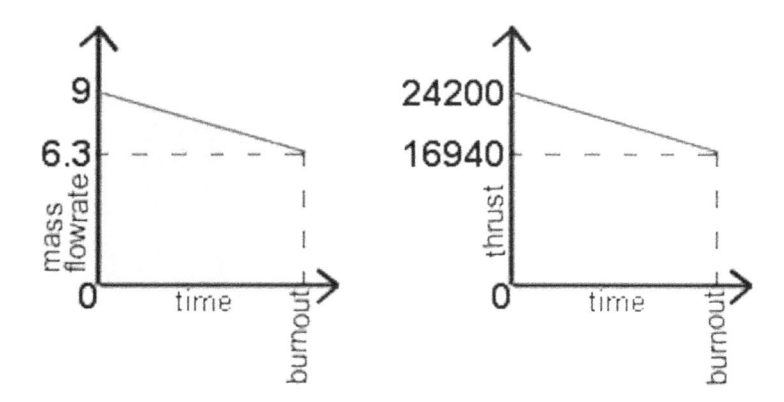

Remember that the nozzle mass flow rate out of the combustion chamber and out the nozzle came from the propellants flowing in, so after every ½ second of sim time we must remember to lower the spacecraft's propellant mass. And we must keep doing this as time goes on until the propellant all runs out at burnout.

But in the graphs above, we don't know the burn time: at what time does burnout occur?

As we noted in equation 9.21 the total propellant mass is the time integral of nozzle mass flow rate.

For the purposes of the sim, we'll assume that the total propellant mass is the X = 310.6 kilograms of propellant we calculated earlier:

$$\text{Total propellant mass} = 310.6 = \int_{start\ time}^{end\ time} \dot{m}_{nozzle}\ dt$$

where the start time is engine ignition and the end time is the time when the liquid Nitrous runs out (when X kilos of propellant has been used up, but there is still the remaining vapour propellant mass).

Now the coloured 'area under' the mass flow rate graph above (which is equal to the integration, i.e. the total propellant mass) is a trapezium: a rectangle with a triangle on top of it. Geometrically, this has a total area equal to the average mass flow rate over the burn times the burn time.

This information lets us calculate the burn time (equal to: burnout – ignition) as:

$$\text{Burn time} = \frac{\text{total propellant mass}}{\text{average mass flow rate over the burn}}$$

$$= \frac{2 \times \text{total propellant mass}}{\text{start mass flowrate} + \text{burnout mass flowrate}} = \frac{2 \times 310.6}{9 + 6.3} = 40.5 \text{ seconds}\ \ \text{(Sim equ.4)}$$

This equation will only work for this linearly sloping mass flow rate curve.

So liquid burnout occurs at 40.5 seconds.

The gradient of the downward slope of the nozzle mass flow rate graph is:

$$\frac{\text{burnout mass flow rate} - \text{start mass flow rate}}{\text{burn time}} = \frac{6.3 - 9}{40.5} = -0.0668\ \ \text{(Sim equ.5)}$$

So after every second we reduce the previous nozzle mass flow rate value by 0.0668 × 1/2 second.

Similarly, the gradient of the thrust curve is:

$$\frac{burnout\ thrust - start\ thrust}{burn\ time} = \frac{16940 - 24200}{40.5} = -179 \quad \text{(Sim equ.6)}$$

After every second we reduce the previous thrust value by 179 × 1/2 second.

Programming in Microsoft Office's Excel

Now it's time to program the Excel spreadsheet. Programming is like following a recipe in a cookbook: unless you follow the recipe exactly then it'll all go wrong. You must put the words and numbers in *exactly* the same rows and columns as I have. (The words are just for you; the computer ignores them and just reads the numbers.)

The menus and symbols at the top may look different depending on what version of Microsoft Office you're using, but it's what is below this on the worksheet that counts.

Each box on the spreadsheet is called a cell: it has a letter at the top of the column to show which column it is in and a number at the left of the row to show what row it is in.

I've written a generic simulation in Excel which is downloadable from the Aspirespace website

www.aspirespace.org.uk/technical_papers.html

Click on 'A rocket trajectory simulator using high school physics'.

To save time and text, we'll modify this sim rather than laboriously code it all again from scratch. Now read the documentation within the download to discover how the sim functions.

[In essence, we integrate acceleration (from Newton's 2nd law) to get velocity, which we integrate again to get height. While doing this we feed in drag and gravity.]

The engine

We have to modify the first worksheet 'Engine' of the sim to insert our engine. Select the 'Engine' worksheet using the tabbed labels and arrows at the very bottom of the sheet.

First, delete the text 'ESTES C6-5 engine etc' from cells A1 and A2 of the worksheet.

We need more empty space at the top of the page. So highlight the first three rows of the worksheet by holding the left mouse button down, and dragging the mouse cursor over the numbers 1,2,3 as shown here:

Then right-click your mouse and choose 'insert' to insert three new empty rows.

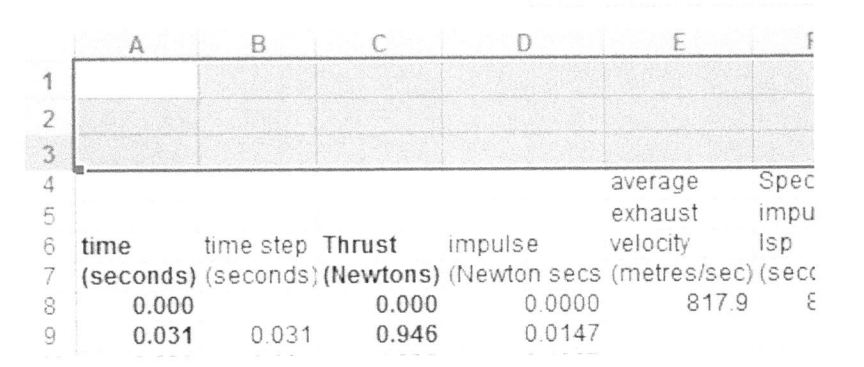

	A	B	C	D	E	F
1						
2						
3						
4					average	Spec
5					exhaust	impu
6	time	time step	Thrust	impulse	velocity	Isp
7	(seconds)	(seconds)	(Newtons)	(Newton secs)	(metres/sec)	(secc
8	0.000		0.000	0.0000	817.9	8
9	0.031	0.031	0.946	0.0147		

119

Now do exactly the same again to add another three empty rows.

Now, for tidiness, delete the block of cells between cell D10 and F34 inclusive as they're not required (click cell D10 and then - whilst holding down the left mouse button - move the mouse cursor from cell D10 to cell F34, then hit the 'delete' key). Also delete the contents (delete key) of cell G35.

Next, type in the following words and start numbers until your Excel page looks *exactly* like this, and in the positions indicated:

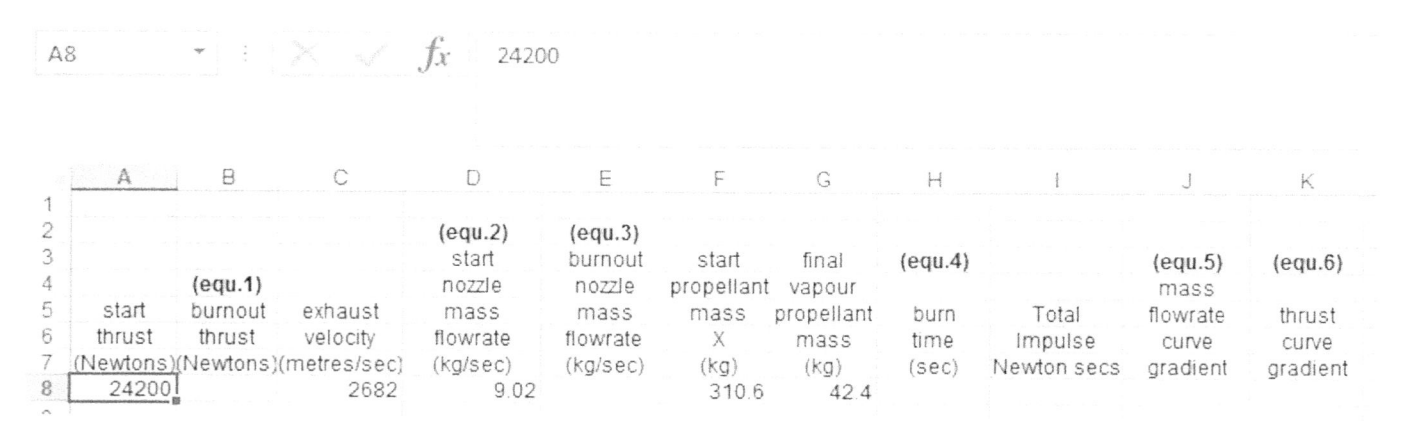

	A	B	C	D	E	F	G	H	I	J	K
1											
2				(equ.2)	(equ.3)						
3				start	burnout	start	final	(equ.4)		(equ.5)	(equ.6)
4		(equ.1)		nozzle	nozzle	propellant	vapour			mass	
5	start	burnout	exhaust	mass	mass	mass	propellant	burn	Total	flowrate	thrust
6	thrust	thrust	velocity	flowrate	flowrate	X	mass	time	Impulse	curve	curve
7	(Newtons)	(Newtons)	(metres/sec)	(kg/sec)	(kg/sec)	(kg)	(kg)	(sec)	Newton secs	gradient	gradient
8	24200		2682		9.02		310.6	42.4			

What we're going to do next is fill row 8 with data <u>concerning the liquid phase part of the burn only</u>.

Now move your mouse over cell B8 (column B, row 8) and left-click it. A box will appear round the cell as is shown in the picture below. In the little formula window at the top of the sheet you can see the text '=A8*0.7'

That's exactly what you must type in the cell, and you *must* include the '=' sign to tell the computer that what you're typing in is a formula rather than text. Typing this formula in tells the computer to multiply whatever number is in cell A8 (the start thrust) by 0.7 and fill cell B8 (the burnout thrust) with the answer. The '*' sign is the multiply sign that computers use instead of '×'.

Don't type in the ' ' signs, I've just added those to tell you to type-in what's between them.

This sum gives us the burnout thrust. (We named this simulation equation 1 earlier on page 117: Sim equ.1)

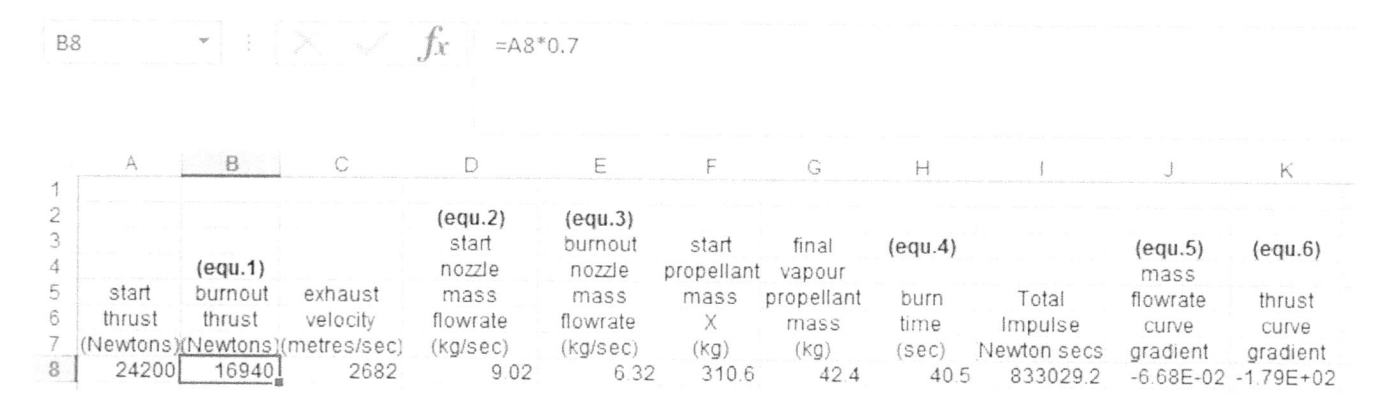

B8 f_x =A8*0.7

	A	B	C	D	E	F	G	H	I	J	K
1											
2				(equ.2)	(equ.3)						
3				start	burnout	start	final	(equ.4)		(equ.5)	(equ.6)
4		(equ.1)		nozzle	nozzle	propellant	vapour			mass	
5	start	burnout	exhaust	mass	mass	mass	propellant	burn	Total	flowrate	thrust
6	thrust	thrust	velocity	flowrate	flowrate	X	mass	time	Impulse	curve	curve
7	(Newtons)	(Newtons)	(metres/sec)	(kg/sec)	(kg/sec)	(kg)	(kg)	(sec)	Newton secs	gradient	gradient
8	24200	16940	2682		9.02	6.32	310.6	42.4	40.5	833029.2	-6.68E-02	-1.79E+02

Now click on cell D8 and type in '=A8/C8' which means take the number in cell A8 (the start thrust) and divide it by the number in cell C8 (the effective exhaust velocity). The '/' sign means 'divide by'. This gives us the nozzle mass flow rate at the start (we called this Sim equ.2)

Click on cell E8 and type in '=D8*0.7'

This gives us the nozzle mass flow rate at the end (Sim equ.3). Remember that it was the start value times 0.7

Click on cell H8 and type in '=2*F8/(D8+E8)' This gives the burn time (Sim equ.4).

The brackets ( ) tell the computer to add D8 and E8 together first before doing the rest of the sum. You'd get a different answer without the brackets.

Now we can work out the rocket's Total impulse.

Remember from equation 9.20 that the Total impulse is the time integral of thrust, which is of course equal to the area under the thrust curve. A little algebra says that this area is equal to the average thrust times the burn time: '=0.5*(A8+B8)*H8'   Type this into cell I8.

Click on cell J8 and type in '=(E8-D8)/H8'

This gives the gradient of the nozzle mass flow rate curve (Sim equ.5).

Finally, click on cell K8 and type in '=(B8-A8)/H8'

This gives the gradient of the thrust curve (Sim equ.6).

Your spreadsheet should now look exactly like the picture towards the bottom of the previous page.

I've widened some of the columns and narrowed others. To do this, put your cursor between two columns at the top of them and the cursor will turn into a little cross shape:

Then holding down the left mouse button, drag the column to the required width. The width won't affect the sim in any way, it's just for you to read.

If a cell is not displaying the figures to the required number of decimal places, right mouse button-click it, select 'format cells' then 'number' and choose how many decimal places to display.

The thrust and nozzle mass flow rate curves

The next step is to program-in the thrust and mass flow rate graph curves.

First, we need to divide the curves into steps; twenty steps is sufficient. In cell A15 type in '=A14+H$8/20'

Put your mouse cursor over the little square in the bottom right-hand corner of this cell, and keeping the left-hand mouse button clicked, drag this cell down to cell A34. This will also fill cells A16 to A34 with values.

The '$' sign is a code to tell Excel always to use the value in row 8 of column H, which is the burn time.

Now for the thrust curve: click on cell C14 and type '=A$8+K$8*A14'

As you did for column A, drag this cell down to row 34. What we're doing is using the gradient of the thrust graph curve which we got from equation 6 to calculate the thrust at each time step.

In cell C35 type in '0' (zero).

Similarly, in cell G14, type '=D$8+J$8*A14' and drag this cell down to row 34. You'll see that this is filling the column with the nozzle mass flow rate graph curve.

The mass curve

In column H, numerical integration is used to work out how much mass has gone out the nozzle after every time-step in column B.

$$\text{Decrease in propellant mass} = \int_0^{time\ step\ size} \dot{m}_{nozzle}\, dt \quad \text{equ. 9.29}$$

though first we have to input the start propellant mass (including vapour): in cell H14 type in '=F8+G8'.

Cells H15 to H34 perform this integration using the nozzle mass flow rate data in row G.

In cell H35 type in '0' (zero).

The vapour burn

Now that we've calculated the liquid phase part of the burn, we need to add the effect of the vapour burn after the liquid Nitrous has run out.

As we noted earlier, the Total impulse (TI) from the vapour burn is roughly 13% of the total impulse of the entire burn (liquid plus vapour):

$$TI_{vapour} = 0.13(TI_{liquid} + TI_{vapour}) \qquad \text{equ. 9.30}$$

rearranging:

$$(1 - 0.13)TI_{vapour} = 0.13(TI_{liquid}) \qquad \text{equ. 9.31}$$

giving:

$$TI_{vapour} = \frac{0.13}{(1-0.13)}(TI_{liquid}) = 0.15(TI_{liquid}) \qquad \text{equ. 9.32}$$

so in cell D35 type in '=I8*0.15' (Recall that cell I8 is the Total impulse of the liquid phase alone.)

In cell E35 type in 'vapour total impulse'. You can highlight cell D35 in yellow if you like.

Now that we have the vapour Total impulse, we need to add the vapour burn to the end of the thrust curve in such a way that the vapour Total impulse is as in cell D35. For simplicity, we'll model the vapour burn as a triangle shape on the end of the liquid thrust curve graph, the slope of which descends to zero. The area under/of this triangle is, from integration, equal to the vapour Total impulse.

The area of a triangle is half times the base times the height. So:

$$\tfrac{1}{2} \times burn\_time_{vapour} \times height = TI_{vapour} \qquad \text{equ. 9.33}$$

rearranging:

$$burn\_time_{vapour} = \frac{2 \times TI_{vapour}}{height} \qquad \text{equ. 9.34}$$

'height' is simply the last entry in the liquid thrust curve (cell C34), which we dock the triangle onto.

Type this vapour burn time into cell B35: '=2*D35/C34' which gives 14.8 seconds. Now add this to the time of cell C34 by typing in cell A35: '=A34+B35' which gives 55.25 seconds.

This completes the changes to the 'Engine' worksheet, our hybrid engine is programmed-in.

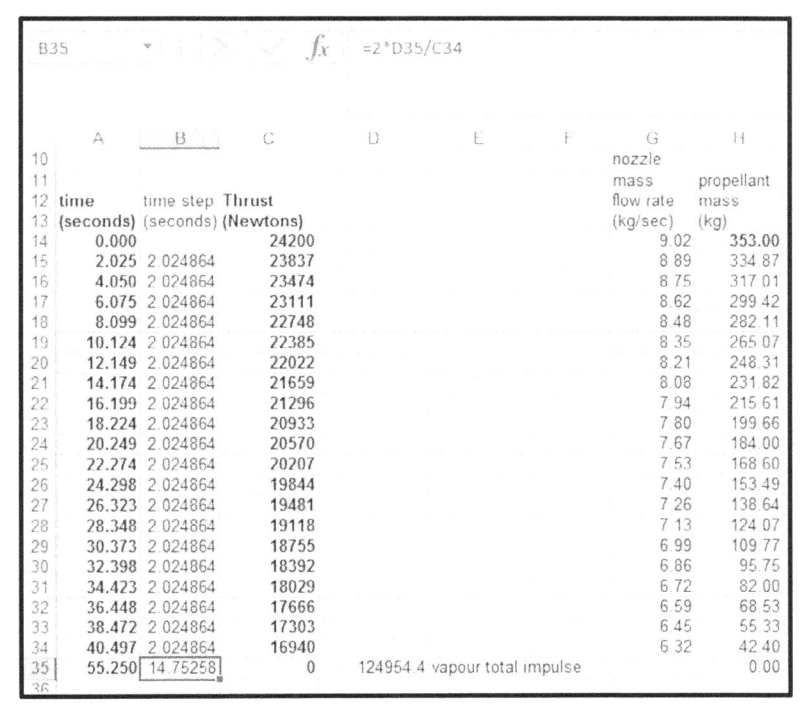

As a program check, check that columns O, P, and Q, have the entries shown here:

time (seconds)	Thrust (Newtons)	Propellant mass (kg)
0.000	24200.0	353.0
2.025	23837.0	334.9
4.050	23474.0	317.0
6.075	23111.0	299.4
8.099	22748.0	282.1
10.124	22385.0	265.1
12.149	22022.0	248.3
14.174	21659.0	231.8
16.199	21296.0	215.6
18.224	20933.0	199.7
20.249	20570.0	184.0
22.274	20207.0	168.6
24.298	19844.0	153.5
26.323	19481.0	138.6
28.348	19118.0	124.1
30.373	18755.0	109.8
32.398	18392.0	95.8
34.423	18029.0	82.0
36.448	17666.0	68.5
38.472	17303.0	55.3
40.497	16940.0	42.4
55.250	0.0	0.0

[If you're mathematically astute, you might object to the inherent simplification that the propellant mass drops linearly over the vapour burn time (cells H34 to H35). Feel free to calculate the actual mass curve and feed it in by inserting additional rows after row 34 to split the 14.8 seconds of the vapour burn into smaller chunks of time. You will then have to increase the range of the linear interpolation function in column D of the 'Sim' worksheet.]

The sim worksheet

Now that the engine is programmed-in, we can fire up the simulation in earnest. Select the 'Sim' worksheet tab at the bottom of the spreadsheet.

In the user-inputs box (row 10) we input the sim initial conditions.

- In cell B10, set the time-step to 0.5 seconds.
- In cell C10, set the empty mass to 400 kilograms.
- In cell D10, set the fuselage diameter. I estimate that our craft needs to be 1.54 metres wide to fit you into it. This gives a cross-sectional area of 1.863 square metres (as calculated in cell G10) as is shown in dark grey here:
- In cell E10, set the start altitude to 21336 metres (70,000 feet).

We need more engine data: click on cell C54, then hold down the left mouse button and drag across both of cells C54 and D54 to draw a box around both cells. Then drag both cells down to row 127. In cells C128 and D128 type in '0' (zero).

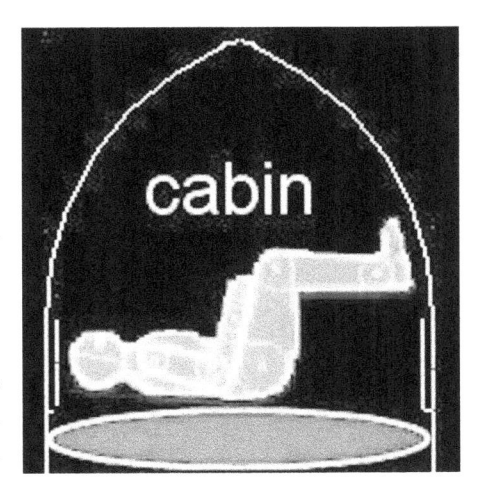

Now we select which drag data to use, 'power on' (engine thrusting) or 'power off' (engine post-burnout): drag cell Q54 down to cell Q127 because row 127 is just after burnout.

Next, we need more sim data. Highlight (hold down left mouse button and drag across) <u>the entirety</u> of row 186 (cells B186 to T186) and drag <u>the entire row</u> far down to row 337 which is where the velocity in row H337 goes negative, which signifies that apogee has just been passed, and the vehicle is now moving downwards. You may have to widen column I to display the large values of altitude otherwise you might see cell message '######' displayed.

The sim results

Apogee (the highest height value) should be 105,819 metres (106 Kilometres) in cell I336.

Look at the acceleration-in-gees column G: the numbers increase to 3.6 gees at liquid phase burnout (though see below). This is less than 4 gees so we're okay. That's why I picked the start thrust that I did, so that the gees wouldn't get too high. (I had to run the sim several times till I got the right start thrust.)

Accuracy

How accurate is this simple sim? Well one way to find out is to reduce the time-step. Try a time-step of 0.25 seconds in cell B10: the apogee only reduces by 1 kilometre, that's pretty good, there's little point in reducing the time-step further.

Sims do get more accurate with smaller time-steps, but their overall accuracy is only as accurate as the thrust curves and drag coefficients (chapter 1) that are put in. Our thrust curve and drag coefficient aren't accurate enough that a smaller time-step would make a difference to the overall accuracy.

[If the timestep is *really* small, (millionths of a second) this causes tiny changes in the sim numbers. The changes are so tiny that the computer can't store these numbers properly, and the accuracy drops. This is **rounding error**.]

A more detailed sim

I have at home a much more detailed simulation than the one we've just done. I've written it in the Microsoft Visual C++ programming language and it simulates the hybrid engine in a lot more detail. In particular, it models the vapour burn correctly, and doesn't make our approximation of a constant effective exhaust velocity during the liquid burn.

I burned exactly the same amount of propellant, 353 kilograms, and used the same drag and atmosphere data.

The result of my sim is an apogee of 105 kilometres compared to the Excel's 106 kilometres. This error is well within the range of accuracy of the drag data and vapour burn simplifications.

Also, the Excel sim assumes an imaginary nozzle that is always ideally expanded (no pressure thrust, positive or negative) whereas my sim uses a real nozzle: if I switch to an ideal nozzle I gain an extra kilometre which brings the results closer.

Of course we're only comparing one sim with another, rather than comparing sim with actual flight data, which is why I arranged the sims to exceed 100 kilometres by a margin of error of around 5 kilometres, just to be on the safe side.

In my sim, the liquid phase burnout was four gees as required.

The Nitrous run-tank

The trajectory sims say that 353 kilograms of propellant will get us into Space.

The best mixture (halfway between the frozen and shifting results) for Nitrous and high-density polyethylene (HDPE) is 7 kilograms of Nitrous for every 1 kilogram of HDPE fuel. Using this mixture ratio of 7 gives:

$$\textbf{Nitrous mass} = 353 \left(\frac{7}{7+1}\right) = \textbf{309 kilograms} \qquad \text{equ. 9.35}$$

Assuming that we use 12 percent of the tank as ullage (see pages 35 and 36), 309 kilograms of Nitrous takes up a volume of:

$$(1 + 0.12)\left(\frac{309}{786.6}\right) = 0.434 \text{ cubic metres } (= 434 \text{ litres}) \qquad \text{equ. 9.36}$$

(assuming a liquid Nitrous density of 786.6 kg/m³ at 20 degrees C, see chapter 2).

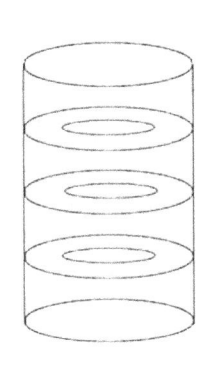

We now have the size of our Nitrous run-tank, a sphere of 434 litres as drawn here, which is surprisingly small because spheres have the largest internal volume for their diameter:

However, this run-tank is large enough that we'll need to fit **slosh baffles** into the tank to reduce the slosh mass (the mass of liquid that is sloshing to and fro across the tank see page 33) otherwise the control system will get upset.

These slosh baffles are metal hoops (coloured here) that are stuck to the walls of the tank. They physically break up the waves that slosh from side to side, so there's less mass sloshing about.

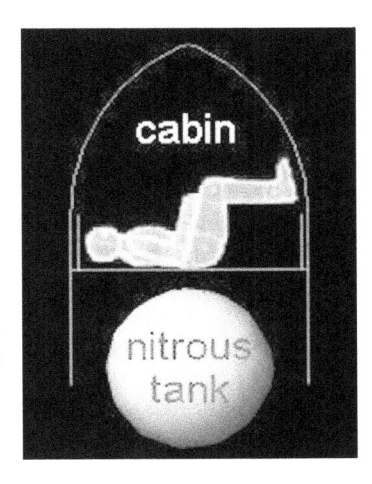

The injector

We now have to design the injector that sprays the Nitrous into the combustion chamber akin to a shower head.

I know from my experience with Nitrous hybrids that a good size to make each of the many holes in the Nitrous injector plate is 1.5 mm wide. Don't make the orifices larger than this or you won't get proper atomization (see below) but smaller orifices work better.

There are many different shapes of orifice we could make, but for Nitrous a simple plain drilled hole does adequately.

This picture shows a simple showerhead, which is all right for smaller engines, but for our larger engine we want to make sure the Nitrous is properly **atomised**: broken into small enough droplets of liquid Nitrous so that the droplets get heated into a vapour easily.

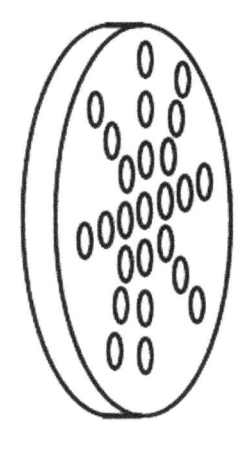

[A smaller droplet has a relatively larger surface area for the mass of Nitrous it contains, and it's the surface area compared to the mass of the droplet that determines how quickly it gets vaporized by the hot combustion chamber gasses.]

To atomise the liquid properly, we'll arrange for the streams of liquid coming out of the orifices to splash into each other. We'll splash together pairs of streams; this is called a **doublet injector**.

Alternatively, you could splash the streams into a wall, which is called a **splashplate injector**.

Now we have to choose how many orifices to make.

The mass flow rate of Nitrous through an injector depends on the pressure drop between both sides of it (ΔP = tank pressure minus chamber pressure) and the total cross-sectional area of all the orifices added together: the more holes, the greater the mass flow rate that will pass through the injector (see the injector equation in appendix 3).

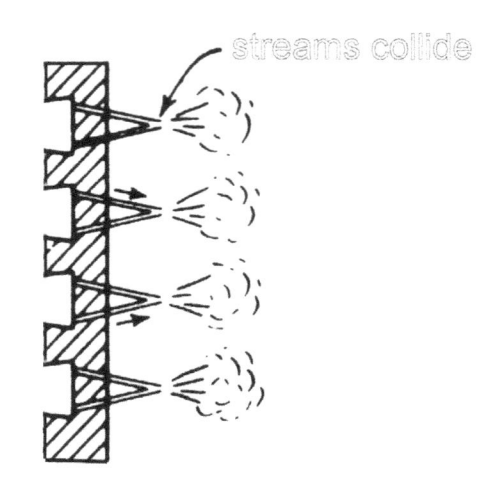

What Nitrous flow rate is required?

Shortly after the start of the burn we know that in the chamber:

$\dot{m}_{out} = \dot{m}_{in}$ and that $\dot{m}_{in} = \dot{m}_{fuel} + \dot{m}_{nitrous}$

We could design the injector to use the Nitrous flow rates at the start of the burn, but when you're designing hybrids, it's better to use the values that you get halfway through the liquid-Nitrous-phase burn as this is a handy average, and one should design a hybrid to be burning stoichiometrically around liquid mid-burn.

For the right mixture we need to burn 7 parts Nitrous to 1 part of fuel, so we want a Nitrous mass flow rate of 7/8ths of the total $\dot{m}_{in}$, where $\dot{m}_{in}$ is equal to the mid-liquid-burn nozzle mass flow rate, which we can work out as the average of the start and end flow rates - as given on our earlier spreadsheet engine page cells G14 and G34 - because the mass flow rate graph is a straight line:

$$\frac{9.02+6.32}{2} = 7.67 \text{ kilograms per second out} \quad \text{equ. 9. 37}$$

So we need $\frac{7}{8} \times 7.67 = 6.71$ equ. 9. 38

kilograms per second of Nitrous to pass through the injector.

For suitable tank and chamber pressures to get at the liquid phase mid-burn (see Appendix 2),

ΔP = tank pressure - chamber pressure = 10 Bar (1,000,000 Pascals)   equ. 9. 39

Using the injector equation App. 3. 14 given in appendix 3 and the liquid phase mid-burn values from Appendix 2:

$$N = \frac{\dot{m}_{liquid}}{A_{orifice}\sqrt{\frac{2\rho_{liquid}\Delta P}{C}}} = \frac{6.71}{\left(\frac{\pi \times 0.0015^2}{4}\right)\sqrt{\frac{2 \times 835 \times 1,000,000}{2.0}}} = 132 \text{ orifices to be drilled in total}$$

equ. 9. 40

The combustion chamber

The trajectory sims say that 353 kilograms of propellant will get us into Space, of which 309 kilograms is Nitrous.

So the:

HDPE mass = 353 – 309 = 44 kilograms   equ. 9. 41

As the engine is large we'll need four ports (see chapter 7), and we'll use four combustion chambers for simplicity. Having more than one chamber weighs a bit more than one chamber with four ports inside it, but it's a lot easier to make.

We'll have four chambers with $\frac{44}{4} = \mathbf{11}$ **kilos** of fuel in each   equ. 9. 42

but instead of four nozzles we'll use one common nozzle (to save money) of the same size as above. The four chambers will feed into it.

So our spacecraft will look like this:

The large nozzle is the one with the area ratio of 70 that we chose in the previous chapters, designed as a bell nozzle using appendix 6.

We can put a smaller nozzle on just one of these chambers for sea-level testing.

Combustion chamber design

Now to design the combustion chambers geometry.

We have four combustion chambers exiting into a common nozzle. At the start of the burn, each combustion chamber is handling a mass flow rate of (see equation 9.37):

$$\frac{9.02}{4} = \mathbf{2.255} \text{ kilograms per second Nitrous flow rate} \quad \text{equ. 9. 43}$$

Port diameter

As the combustion chamber reaches operating pressure (start pressure, Appendix 2) we want a maximum flux leaving the end of the port of around 870 kg/s m². From chapter 1, this sizes our initial port cross-sectional area:

$$A_{port\_initial} = \frac{\dot{m}_{port}}{G_{port}} = \frac{2.255}{870} = \mathbf{0.00259} \text{ m}^2 \quad \text{equ. 9. 44}$$

We get this from a port diameter of:

$$D = \sqrt{\frac{4\, A_{port\_initial}}{\pi}} = \mathbf{0.0574} \text{ metres = 57.4 millimetres} \quad \text{equ. 9. 45}$$

The final port diameter depends on the exact regression rate relationship for your fuel, and of course the fact that a Nitrous hybrid oxidiser mass flux into the port drops with time as Nitrous tank pressure drops, which makes the analysis complicated.

However, as a guide, I said on page 92 that the average regression rate over the liquid Nitrous part of the burn was around 0.8 millimetres per second for Nitrous and HDPE at typical port mass fluxes and a large size.

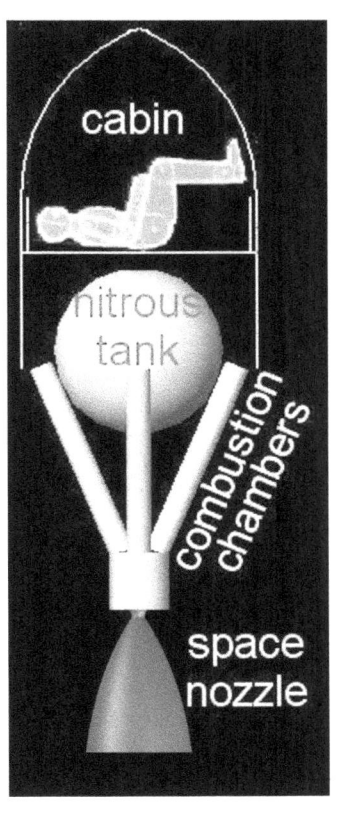

We know from earlier in this chapter that the total liquid burn time is 40.5 seconds, so the final port diameter at liquid burnout is roughly:

$$57.4 + (40.5 \times 0.8) \times 2 = 122 \text{ millimetres} \qquad \text{equ. 9. 46}$$

At the mid-burn of the liquid phase (a result we'll use shortly) the port diameter will be:

$$\frac{57.4 + 122}{2} = 89.7 \text{ millimetres (0.0897 metres)} \qquad \text{equ. 9. 47}$$

On page 40 I noted that the burn time of the vapour phase is around 67% of the liquid phase, giving a vapour phase burn time of:

$$40.5 \times 0.67 = 27.1 \text{ seconds} \qquad \text{equ. 9. 48}$$

From page 92, the average regression rate over the vapour Nitrous part of the burn is around 0.2 millimetres per second for Nitrous and HDPE at typical port mass fluxes and a large size.

This adds extra to the final port diameter:

$$122 + (27.1 \times 0.2) \times 2 = 133 \text{ millimetres} \qquad \text{equ. 9. 49}$$

So our combustion chamber ablative liner should have an inner diameter of 133 millimetres if we want to burn out all the fuel. If this were a single-chamber engine, we would add a post-combustion chamber (page 86) of 133 millimetres length after the fuel grain.

<u>Port length</u>

Now we need to estimate the port length that will give us stoichiometric burning at around mid-burn of the liquid phase. Our liquid-burn-averaged regression rate of 0.8 millimetres per second (0.0008 metres per second) is a reasonable guide, as the average value will occur near the liquid mid-burn.

The fuel mass flow rate at this point depends on the regression rate and the surface area S of the fuel port:

$$\dot{m}_{fuel} = \dot{r}\, \rho_{fuel}\, S = \dot{r}\, \rho_{fuel}\, (\pi D L) \qquad \text{equ. 9. 50}$$

where D is port diameter and L is port length (this equation assumes a slow regression rate).

For stoichiometry, we need to burn 7 parts Nitrous to 1 part fuel, so the fuel mass flowrate is 1/8th of the total chamber mass flow rate, which itself is equal to ¼ of the nozzle mass flow rate as there are 4 combustion chambers exiting into one common nozzle.

$$\frac{\dot{m}_{nozzle}}{8 \times 4} = \dot{m}_{fuel} = \dot{r}\, \rho_{fuel}\, (\pi D L) \qquad \text{equ. 9. 51}$$

and from my graph on page 38 we know that the nozzle mass flow rate during the liquid phase drops pretty well linearly with time.

So at the liquid phase mid-burn, it should be halfway (the average) between its start value, and its liquid phase burnout value as given by Sim equ.3 earlier:

$$\textbf{Average liquid phase } \dot{m} = \frac{9 + 6.3}{2} = \mathbf{7.65} \text{ kg/s} \qquad \text{equ. 9. 52}$$

Rearranging the above equation to get port length:

$$L = \frac{\dot{m}_{nozzle}}{8 \times 4 \times \dot{r}\, \rho_{fuel}\, \pi D} = \frac{7.65}{32 \times 0.0008 \times 935 \times \pi \times 0.0897} = 1.1 \text{ metres} \qquad \text{equ. 9. 53}$$

where the density of my preferred brand of HDPE fuel is 935 kg/m³, and the mid-burn port diameter is 0.0897 from equation 9.47

Do bear in mind that this is just an estimate. Tuning of the port length to achieve stoichiometry has to come from actual test-firing data. As it's much easier to reduce a combustion chamber length after it's built rather than increase it, then make your fuel grain over-long until you've got the test data to tune it.

Combustion chamber simulation

Rules of thumb and average regression rate estimates are all very well for a first ballpark design, however once accurate static test data is available, a more accurate method is required.

'Humble' chapter 7 section 7.5.2 (see references) describes a simulation one can code up to model the regression more accurately, which involves simulating the entire engine. We at Aspirespace coupled this simulation to the simulation of the Nitrous run-tank emptying described at the end of chapter 3 of this book.

The core component of the sim is modelling the change of combustion chamber mass with time in a similar manner to equation 4.17 of chapter 4. But with a hybrid, the combustion chamber volume is increasing with time as the fuel grain erodes away with time, and this has to be dealt with.

The rate at which the fuel port surface erodes is:

$$\dot{r}_{average} S \quad \text{equ. 9.54}$$

where $S = 2\pi \, r_{average} \, L$ is the port surface area (assuming a circular port).

The chain rule of differentiation shows that this term is also equal to the time rate of change of chamber volume, or the *volumetric* flow rate (mass flow rate divided by gas density) of gasified fuel.

$$\frac{dV}{dt} = \frac{dV}{dr}\frac{dr}{dt} = \frac{d}{dr}\left(\pi \, r_{average}^2 \, L\right) \dot{r}_{average} = 2\pi \, r_{average} \, L \, \dot{r}_{average} = \dot{r}_{average} \, S$$
equ. 9.55

Mass accumulation within the chamber with time - resulting from the imbalance of gas inflow minus gas outflow - can then be modelled using the following unsteady lumped-parameter method of Heister and Landsbaum.

The ideal gas law can be rearranged to yield the current mass of gas within the combustion chamber.

From equation 1.15:

$$m = \frac{PV}{RT} \quad \text{equ. 9.56}$$

Taking the natural logarithm of both sides of this equation, and using the law of logarithms:

$$\ln(m) = \ln\left(\frac{PV}{RT}\right) = \ln(P) + \ln(V) - \ln(R) - \ln(T) \quad \text{equ. 9.57}$$

differentiating this using the chain rule, one obtains:

$$\frac{d}{dm}(\ln(m))\frac{dm}{dt} = \frac{d}{dp}(\ln(P))\frac{dp}{dt} + \frac{d}{dV}(\ln(V))\frac{dV}{dt} - \frac{d}{dR}(\ln(R))\frac{R}{dt} - \frac{d}{dT}(\ln(T))\frac{dT}{dt}$$
equ. 9.58

giving:

$$\frac{1}{m}\frac{dm}{dt} = \frac{1}{p}\frac{dp}{dt} + \frac{1}{V}\frac{dV}{dt} \quad \text{equ. 9.59}$$

(because the derivative of $ln(x) = \frac{1}{x}$ and neither T nor R change with time: their derivative with time is zero).

Rearranging:

$$\frac{dP}{dt} = P\left(\frac{1}{m}\frac{dm}{dt} - \frac{1}{V}\frac{dV}{dt}\right) \quad \text{equ. 9.60}$$

where $\frac{dm}{dt}$ is from equation 4.14, chapter 4, and $\frac{dV}{dt}$ is from equation 9.55 above. ($\frac{1}{m}$ refers to the current gas mass within the combustion chamber, and $\frac{1}{V}$ refers to the current chamber volume.)

This rate of change of combustion chamber pressure is integrated to yield the current chamber pressure at the end of each time iteration. The chamber pressure then feeds back to the tank emptying model discussed at the end of chapter 3.

The engine simulation then proceeds thusly (next page) :

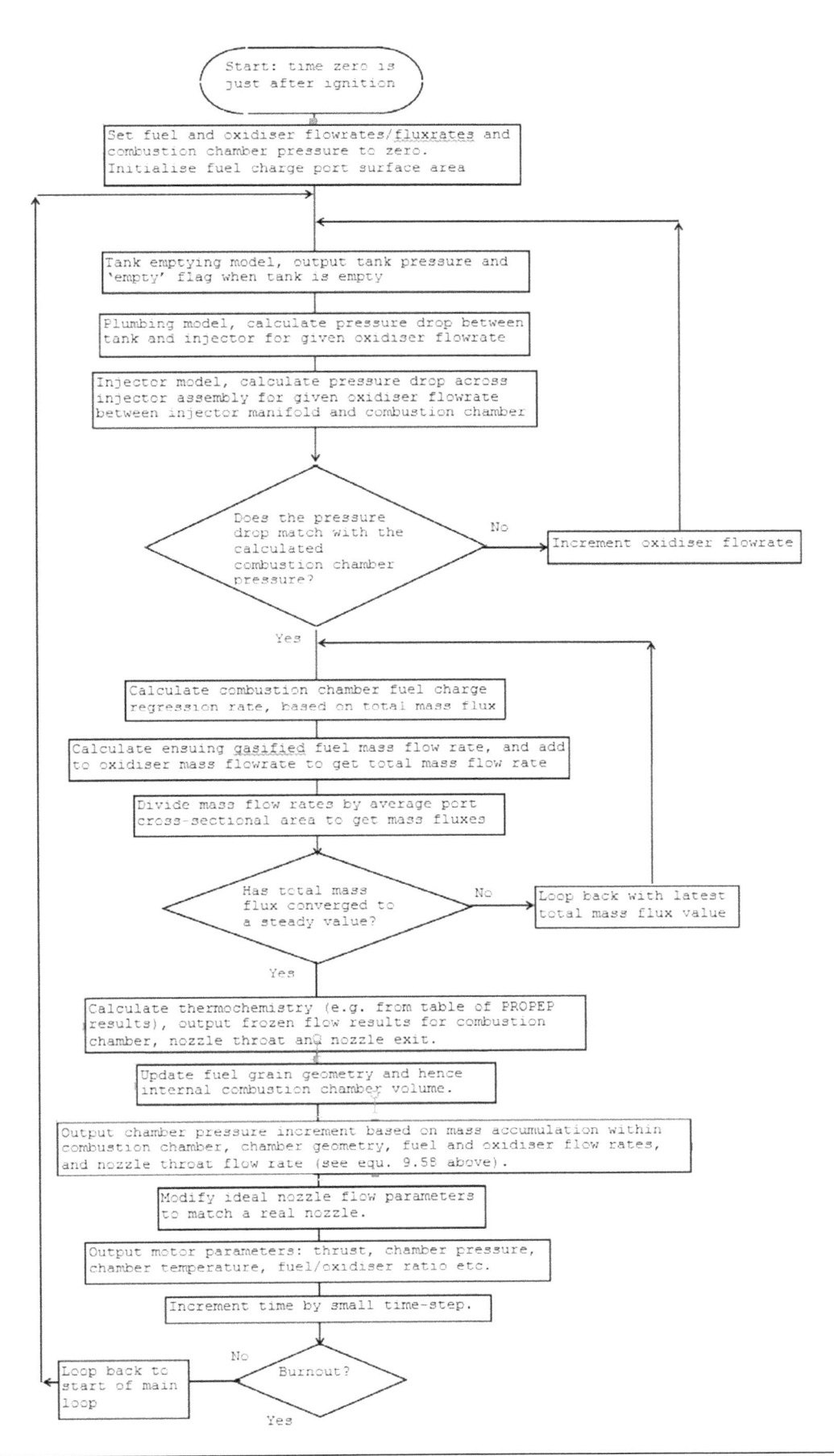

```
┌─────────────────────────┐
│  Start: time zero is    │
│  just after ignition    │
└───────────┬─────────────┘
            ▼
┌──────────────────────────────────────────────────┐
│ Set fuel and oxidiser flowrates/fluxrates and    │
│ combustion chamber pressure to zero.             │
│ Initialise fuel charge port surface area         │
└───────────┬──────────────────────────────────────┘
            ▼
┌──────────────────────────────────────────────────┐
│ Tank emptying model, output tank pressure and    │
│ 'empty' flag when tank is empty                  │
└───────────┬──────────────────────────────────────┘
            ▼
┌──────────────────────────────────────────────────┐
│ Plumbing model, calculate pressure drop between  │
│ tank and injector for given oxidiser flowrate    │
└───────────┬──────────────────────────────────────┘
            ▼
┌──────────────────────────────────────────────────┐
│ Injector model, calculate pressure drop across   │
│ injector assembly for given oxidiser flowrate    │
│ between injector manifold and combustion chamber │
└───────────┬──────────────────────────────────────┘
            ▼
        ╱─────────╲
       ╱ Does the  ╲      No    ┌──────────────────────────┐
      ╱ pressure    ╲──────────▶│ Increment oxidiser       │
      ╲ drop match   ╱          │ flowrate                 │
       ╲ ...       ╱            └──────────────────────────┘
        ╲─────────╱
            │ Yes
            ▼
┌──────────────────────────────────────────────────┐
│ Calculate combustion chamber fuel charge         │
│ regression rate, based on total mass flux        │
└───────────┬──────────────────────────────────────┘
            ▼
┌──────────────────────────────────────────────────┐
│ Calculate ensuing gasified fuel mass flow rate,  │
│ and add to oxidiser mass flowrate to get total   │
│ mass flow rate                                   │
└───────────┬──────────────────────────────────────┘
            ▼
┌──────────────────────────────────────────────────┐
│ Divide mass flow rates by average port           │
│ cross-sectional area to get mass fluxes          │
└───────────┬──────────────────────────────────────┘
            ▼
        ╱─────────╲
       ╱ Has total ╲     No    ┌──────────────────────────┐
      ╱ mass flux   ╲─────────▶│ Loop back with latest    │
      ╲ converged    ╱         │ total mass flux value    │
       ╲...        ╱           └──────────────────────────┘
        ╲─────────╱
            │ Yes
            ▼
┌──────────────────────────────────────────────────┐
│ Calculate thermochemistry (e.g. from table of    │
│ PROPEP results), output frozen flow results for  │
│ combustion chamber, nozzle throat and nozzle exit│
└───────────┬──────────────────────────────────────┘
            ▼
┌──────────────────────────────────────────────────┐
│ Update fuel grain geometry and hence internal    │
│ combustion chamber volume.                       │
└───────────┬──────────────────────────────────────┘
            ▼
┌──────────────────────────────────────────────────┐
│ Output chamber pressure increment based on mass  │
│ accumulation within combustion chamber, chamber  │
│ geometry, fuel and oxidiser flow rates, and      │
│ nozzle throat flow rate (see equ. 9.58 above).   │
└───────────┬──────────────────────────────────────┘
            ▼
┌──────────────────────────────────────────────────┐
│ Modify ideal nozzle flow parameters to match a   │
│ real nozzle.                                     │
└───────────┬──────────────────────────────────────┘
            ▼
┌──────────────────────────────────────────────────┐
│ Output motor parameters: thrust, chamber         │
│ pressure, chamber temperature, fuel/oxidiser     │
│ ratio etc.                                       │
└───────────┬──────────────────────────────────────┘
            ▼
┌──────────────────────────────────────────────────┐
│ Increment time by small time-step.               │
└───────────┬──────────────────────────────────────┘
            ▼
         ╱─────╲
        ╱       ╲   No    ┌──────────────────────┐
       ╱ Burnout?╲───────▶│ Loop back to start   │
        ╲       ╱         │ of main loop         │
         ╲─────╱          └──────────────────────┘
            │ Yes
            ▼
```

The nozzle throat

Now to design the nozzle. We need to calculate the nozzle throat cross-sectional area A^*.

We could design the nozzle to use the pressures and flow rates at the start of the burn, but as I said before, when you're designing hybrids, it's better to use the values that you get halfway through the liquid-Nitrous-phase burn as this is a handy average, and one should design a hybrid to be burning stoichiometrically around liquid mid-burn.

At liquid mid-burn, we want a chamber pressure of around 33 Bar. This gives a reasonable pressure drop across the injector at the end of the liquid phase to avoid terminal flatulence (page 101).

PROPEP 3 results (see appendix 2) tells us that to achieve a chamber pressure of 33 Bar, we need 471 square millimetres of throat cross-sectional area for every *one* kilogram per second of nozzle mass flow rate that passes through the nozzle.

Earlier, we calculated a mid-burn nozzle mass flow rate of 7.65 kilograms per second (equation 9.52), so we multiply this throat area by the flow rate we need:

$$471 \times 7.65 = 3{,}603 \textbf{ square millimetres} \qquad \textbf{equ. 9.61}$$

So we need a throat cross-sectional area A^* of 3,603 square millimetres. To get this A^* we need to make a throat that is:

$$\sqrt{\frac{4 \times 3{,}603}{\pi}} = \textbf{68 millimetres} \text{ in diameter} \qquad \textbf{equ. 9.62}$$

This is for the large nozzle with all four chambers feeding into it.

For a smaller nozzle that we could use just for testing the one chamber alone, it would need to be half the width $\left(\frac{1}{2}\right)$ to get a quarter of the area $\left(\frac{1}{2}\right)^2$, so it would need a throat of:

$$\frac{68}{2} = \textbf{34 millimetres} \text{ in diameter} \qquad \textbf{equ. 9.63}$$

[As you increase length, area goes up by the length squared.]

For our large Space nozzle (fed by 4 chambers) which has an area ratio of 70, the nozzle exit area would then be:

$$70 \times 3{,}603 \text{ which is 252,210 square millimetres (0.25221 square metres)} \qquad \textbf{equ. 9.64}$$

This requires an exit diameter of:

$$\sqrt{\frac{4 \times 252{,}210}{\pi}} = \textbf{567 millimetres (0.567 metres)} \qquad \textbf{equ. 9.65}$$

Chapter 10: Build the hybrid rocket engine

From the previous chapters, we have almost enough information to build the hybrid and then test it.

The nozzle

We'll make the nozzle out of aluminium with an inner heat-proof lining. At the throat, we'll use a thick heat-protective ring of graphite (if it were thin it would crack as graphite is brittle). Unfortunately, graphite conducts heat particularly well, so we'll need to put an insulating layer between the graphite and the aluminium to keep this heat out of the aluminium otherwise it would soften and melt. As an alternative to graphite, we could use carbon-carbon, as this *doesn't* conduct heat to any degree.

The combustion chambers

Each of the four combustion chambers must have enough volume to hold 11 kilograms of HDPE and they need thermal protection from the combustion chamber gasses (see later) which sizes the chamber diameters.

How thick should the combustion chamber walls be?

The thicker the combustion chamber walls are, the heavier they will be. We want to keep the weight down for the flight chambers, so we only want the walls to be just slightly more than thick enough to withstand the chamber pressure without bursting.

Calculating how thick to make the walls is a branch of engineering called **stress analysis**. The **stress** within the wall of a chamber is equal to the force on it (caused by the chamber pressure) divided by the cross-sectional area of the wall (the area of wall you'd get if you cut the chamber in half - across the chamber). The smaller the cross-sectional area (the thinner the wall) the higher the stress will be.

Stress has this symbol: σ and is measured in the same units as pressure: Pascals. The level of stress tells you how hard the bonds between the atoms of your solid material are working to avoid being pulled apart from each other. The smaller the wall area, the fewer atoms there are to hold the chamber together so each atom has to withstand more force; they can only take so much.

Yielding

Each type of material, such as metal, has a maximum stress it can withstand before it starts to **yield** and then breaks. [Plastics and composites don't have a visible yield point on a stress versus strain graph, but do have equivalent safe maximums.]

The combustion chamber wall will stretch slightly like a spring when the engine is firing (stretch is called **strain** which has the symbol ε) because of the chamber pressure, but after burnout the pressure is gone and so the wall should stop being stretched and return to its original size.

This is called **elastic behavior** because the wall stretches then un-stretches like a piece of elastic. (How much strain the stress causes depends on a number called **Young's modulus** or the **modulus of elasticity** which has the symbol E. E is a very large number for most solids, showing that huge stresses cause only very small strains, a stretch of fractions of a millimetre.

If the stretch (strain) is too great caused by too high a stress, the stretch becomes permanent: the wall will stay stretched after engine burnout. This is called yielding, and it shows that permanent damage has occurred to the wall that can quickly lead to the chamber bursting. What has happened is that large groups of the atoms of the wall have been permanently moved; one row of atoms has slipped sideways past its neighbouring row of atoms.

So yielding is to be avoided. The stress in the chamber wall must be kept below the stress which will cause yielding (called the **yield stress** or **yield strength**).

Working out the combustion chamber's stress

Stress engineers visualise the stress in an object by imagining that they've cut it into two sections (this is called **the method of sections**). They then work out how much stress (force) they'd need to create to keep the two parts together, and this is equal to how much stress is in the wall as it tries to stay together.

For example, one of the main stresses on the combustion chamber is called the **longitudinal stress**. It stops the chamber breaking along its length.

Imagine that you've cut a chamber into two sections:

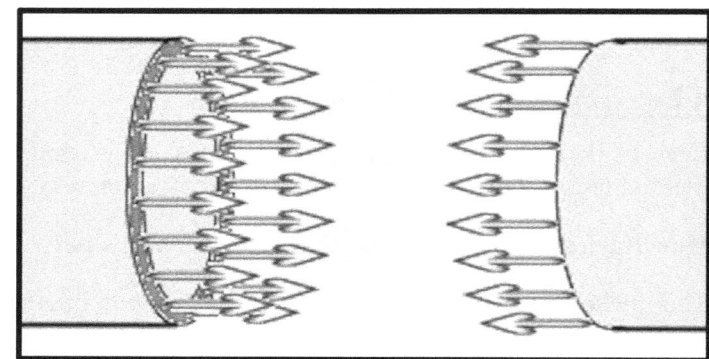

You can then see that the longitudinal stress (the single-ended arrows) trying to hold the two sections together has to exactly equal the force (the double-ended arrows) caused by the chamber pressure that tries to push the two sections apart. (The forces must be exactly equal otherwise one of the sections would begin to move relative to the other.)

The ring-shaped cross-sectional area of wall that the stress has to work with is coloured in dark colour.

The equation to work out the longitudinal stress $\sigma_{longitudinal}$ that the chamber pressure ($P_{chamber}$, in Pascals) causes is simple:

$$\sigma_{longitudinal} = \frac{P_{chamber} \, r_{chamber}}{2t} \qquad \text{equ. 10. 1}$$

where $r_{chamber}$ is the (outer) radius of the chamber in metres, and t is the chamber wall's thickness in metres.

The maximum chamber pressure we expect the rocket to reach is 3,700,000 Pascals (37 Bar) just after ignition (appendix 2), so this makes the stress a very large number.

The other main stress on the chamber is the **hoop stress**. The chamber pressure tries to pull a hoop of the chamber apart, and we can see this by imagining cutting a hoop of the chamber in two:

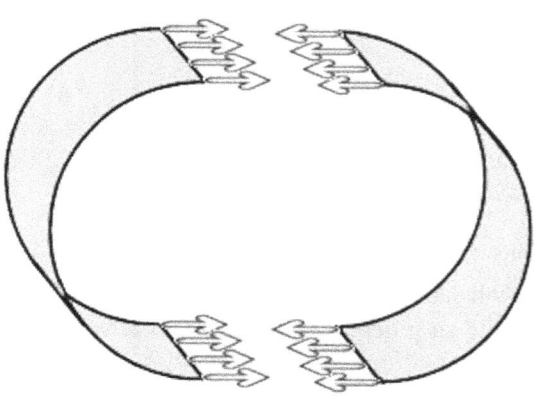

The hoop stress:

$$\sigma_{hoop} = 2 \, \sigma_{longitudinal} \qquad \text{equ. 10. 2}$$

It's exactly double the longitudinal stress.

Because the hoop stress is greater than the longitudinal stress, then
if a tank or combustion chamber bursts it tends to split like a banana skin because it fails in hoop stress first.

We would prefer to simply add the longitudinal stress and the hoop stress together to work out the total stress within the combustion chamber wall. Unfortunately we have to perform a more complex sum than simple addition to combine them because the two stresses are acting in different directions (at right-angles to one another).

One equation for combining the two stresses is the **Von Mises criterion**:

$$\sigma_{total} = \sqrt{\sigma_{hoop}^2 - \sigma_{hoop}\,\sigma_{longitudinal} + \sigma_{longitudinal}^2} \qquad \text{equ. 10.3}$$

But as the hoop stress is double the longitudinal stress, this simplifies to:

$$\sigma_{total} = \sqrt{(2\sigma_{longitudinal})^2 - 2\sigma_{longitudinal}\,\sigma_{longitudinal} + \sigma_{longitudinal}^2} = \sqrt{3}\,\sigma_{longitudinal} \qquad \text{equ. 10.4}$$

Now you might note that $\sqrt{3} = 1.732$ is less than 2, so this equation is saying, rather oddly, that the combined stress is a bit less than the hoop stress, so for safety, we'll design for the hoop stress as this is a more onerous stress condition.

Put the equations above into a spreadsheet and keep changing the wall thickness t until the total stress is just less than the yield stress of whatever material (metal or composite) you're making the chamber wall out of. You can find the yield stress for your material from the internet.

Factor of safety

Actually, you want the total stress - caused by the maximum combustion chamber pressure - to be quite a lot less than the yield stress for safety; you want to include a **safety-factor**. The safety-factor is there because you might not know precisely what the stresses are, they might be higher than you expect.

Also, the material of your wall never yields at exactly the yield stress it says in a book, the actual yield stress could be lower. A sensible safety-factor for something you're going to trust your life with would be at least two (you're intending sitting just above the combustion chambers). This means that your total stress should be no more than half of the yield stress. Yes, this means that the chamber wall will have to be made thicker, which adds more weight, but safety has to come first.

You can't avoid drilling holes into the chamber wall to put bolts or screws through to fix the nozzle and forward end cap to the chamber. These holes raise the total stress locally around the holes: they can raise the average stress by as much as three times. However, provided the holes have bolts in them then the stresses are quite low at this point as the wall is anchored back into the end closure by the bolts, so even with three times the average stress, you won't be anywhere near the hoop stress that you've designed for (assuming you use many small bolts).

It is important to make sure that holes are clean-cut (e.g. reamed-out after being drilled to a smaller diameter) otherwise tiny cracks can occur in the hole wall that will cause failure at lower stress: cracks are notorious stress-raisers.

The combustion chambers walls would be made out of carbon fibre to keep the weight down if this were a flight engine, but for ground testing, we'll make metal chambers with an extra-thick wall for safety (a large safety-factor).

For Aspirespace's combustion chambers, we tend to use 6061 or 6082 seamless aluminium tubes that have been specially heat-treated (called 'T6', the 'T' stands for 'tempering') to give them more strength. They have a yield stress of around 170 MegaPascals (170,000,000 Pascals)

If the chamber's heat insulation (see later) fails anywhere, the metal at that point overheats and loses its heat-treatment. This makes that bit of metal go soft, so it fails by swelling up due to the chamber pressure and popping like a blister. This blister lets us see where the heat leak has occurred without the rest of the chamber splitting or bursting.

Bolt analyses

Just as the chamber will fail if the stress is too high, then the bolts or screws holding the chamber together will also break if they have to carry too much stress.

This diagram shows a section cut through part of the chamber wall and part of the forward end cap, with a bolt holding them together:

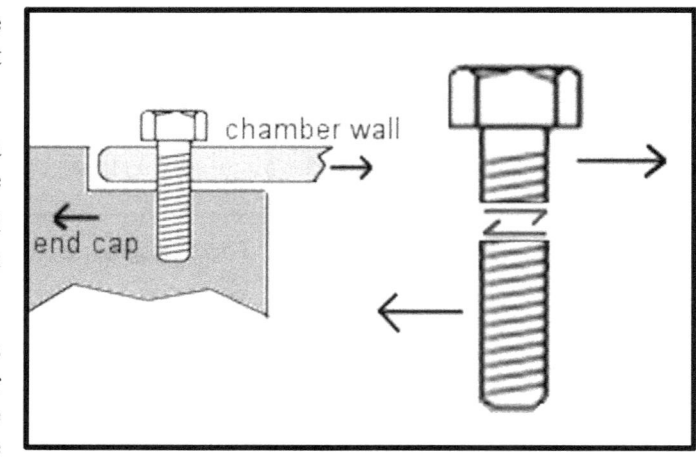

The stress on the bolt is called a **shear stress**, which is a sideways stress: the chamber pressure is trying to tear the bolt across its middle. Shear stress has the symbol τ and is drawn on diagrams as the half-arrows which are drawn on the diagram here in the imaginary gap in the bolt.

If there are 6 bolts in the end cap, then each bolt carries 1/6th of the 'load'; the force caused by the chamber pressure (which is equal to the chamber pressure times the area across the chamber). This 1/6th bolt load is the **factored design strength**. Though don't forget to include your combustion chamber pressure safety-factor: multiply the load by 2.0

The shear stress on each bolt is equal to the force on it divided by the cross-sectional area across the bolt, so the thicker the bolt, the more shear stress it can handle. This shear stress must be less than the shear stress that would cause the bolt to yield, because a bent bolt won't unscrew, and is close to breaking. Again, you can get this maximum allowed shear stress off the internet, or from the bolt manufacturers.

Choose the bolts and screws made from high-strength steel.

Note that it's worth having many smaller bolts rather than a few larger bolts. This is so that the bolts are distributed evenly around the wall of the chamber tube and so can provide a decent seal with an O-ring (see later). Widely-spaced bolts can cause the wall in between bolts to balloon-out and unseal.

Bearing load

However, just because the bolt is strong enough, doesn't mean that the chamber wall can take the load as the bolt bears on it. If the wall is too thin, it'll yield at the bolt hole turning a circular hole into an oval. Worse still, if the bolt hole is too close to the front or rear edge of the combustion chamber tube, it can tear-out: a chunk of tube wall between the hole and the edge is torn right out.

We want to design the bolted shear connections so that the factored design strength - the bearing strength of a single bolt, which equals **0.75 × Rn** - is greater than or equal to the factored design strength.

If $L_C < 2\, d_b$ then $Rn = 1.2\, L_C\, t\, Fu$ equ. 10. 5

If $L_C > 2\, d_b$ then $Rn = 1.4\, d_b\, t\, Fu$ equ. 10. 6

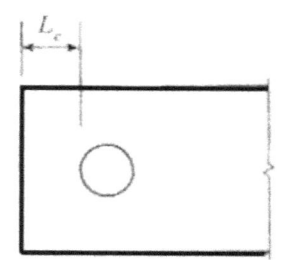

d_b is the bolt diameter, t is the thickness of the wall, Fu is the ultimate strength (stress) of the material and L_C is the distance to the edge of the tube as shown here:

Of course, it's then worth adding a further factor of safety on the bolt load of 1.25.

Nozzle structural fuse

Some designers deliberately choose bolts that are dangerously thin to hold the nozzle on. This is to ensure that the bolts break if the chamber overpressures. Then the whole nozzle flies off to let the chamber gas out just like a blown burst disc, instead of the chamber wall bursting.

This is acceptable for an unmanned rocket vehicle, but for a manned rocket vehicle, when the nozzle bangs off this creates a very large kick of thrust (the combustion chamber pressure times the size of this open hole) that will cause dangerously high gees on you.

The Nitrous run-tank

To build the run-tank for ground testing, we can use a number of thick-walled metal tanks linked together with plumbing.

For the flight tank, we'd want a thin metal liner, perhaps made of two half-spheres welded together. Then we'd wrap composite tape around the liner for strength.

As noted in chapter 8, composite resin and Nitrous don't mix: the resin can decompose the Nitrous, even at moderate temperatures. Composites can often leak under high pressure.

The best option is to use a thin aluminium or fluorocarbon plastic liner (effectively a leak-proof bag) and then wind resin-soaked composite tape around it to make a structure that can take the pressure.

A metal liner will prevent a static electric charge building up. Which metals we choose for the run-tank liner and the plumbing downstream of it is important. Stainless steel or aluminium are compatible, but copper, brass, bronze, and nickel, mustn't be used as they are catalysts for Nitrous oxide (see page 101).

Both aluminium and fluorocarbon plastic (fluoropolymer) need to be welded; you can't make a sphere in one piece. Some types of aluminium lose a lot of their strength when welded, but the older 2000 series aluminiums don't. Welding such thin aluminium requires an expert to do it properly.

Welding fluoropolymer requires it to be done outdoors wearing an oxygen mask. (Keep the heat away from the oxygen.) This is because melted fluoropolymer releases fluorine gas which can make you seriously ill or even kill you even in small amounts if you inhale it. (You can't use any other type of plastic because the rest are all fuels with Nitrous.) Similarly if you decide to 3D print the liner out of, say, FEP fluoropolymer filament, then the air within the printer needs to be fan-extracted up a ventilation shaft.

Run-tank stress analysis

The stress on the spherical Nitrous tank is caused by the Nitrous vapour pressure P_{vapour} which can get as high as 60 Bar (6,000,000 Pascals) on a hot day (see appendix 2).

There's no hoop stress on a pressurised sphere because it isn't hoop-shaped.

There's only the smaller longitudinal stress (imagine slicing the sphere in half) which has exactly the same equation as for the cylindrical combustion chamber:

$$\sigma_{longitudinal} = \frac{P_{vapour}\, r_{sphere}}{2t} \qquad \text{equ. 10.7}$$

Again, we want a Nitrous run-tank with a very large safety-factor (4 or more) when we're testing on the ground, and a safety-factor of at least two for the flight tank (or 1.5 if we use a tank burst disc for protection as we should do).

The injector plates

The injector plates will be made out of metal: either aluminium or stainless steel as they aren't catalysts for Nitrous, and will screw into the metal forward end of the combustion chamber in a way similar to this cut-away sketch:

The metal of the injector plate needs to be fairly thick so that it doesn't bow out or burst due to the pressure (upstream minus downstream) pushing on it.

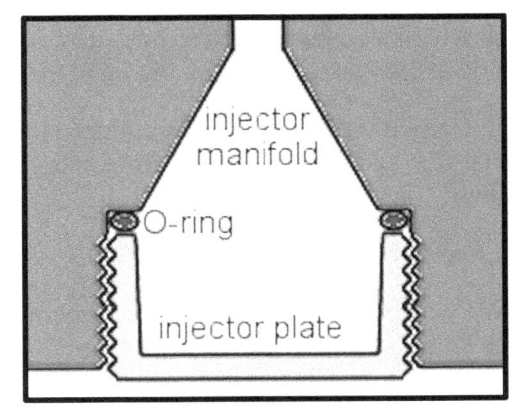

We can calculate how thick the plate needs to be (in metres) using the following formula from Roark's book of stress formulae (see references):

$$thickness = \sqrt{\frac{0.375 \Delta P\, r^3 (1+v)}{\sigma_{max}}} \quad \text{equ. 10.8}$$

where ΔP is the difference in pressure (in Pascals) between one side of the injector and the other, which will be a maximum at ignition when the combustion chamber pressure is that of the atmosphere's pressure at 21.3 kilometres altitude, and the other side is at full initial Nitrous pressure of 51 Bar (see appendix 2).

r is the radius of the plate in metres, and σ_{max} is the maximum (yield) stress the metal of the plate can take (in Pascals), which you can look-up on the web (divide this by a suitable safety-factor of 2).

v is a number called **Poisson's ratio**, that describes how the metal stretches in one direction when pulled in another direction at right angles to it. You can look-up values of v for the metal you're using on the web.

[Note that the diameter of the plate (or rather, the outer ring of orifices drilled in it) has to be less than the initial diameter of the fuel port at ignition, unless there is empty space between the injector and the start of the fuel grain.]

Now we must connect the injectors to the run-tank, or run-tank liner, with electrical wire (unless they're already connected by metal plumbing). This is because as the Nitrous flows through the injectors it can charge up with static electricity.

This could cause an electrical spark which could decompose the Nitrous upstream of an injector and overpressure the plumbing or even the run-tank. The wires keep the different metal components of the rocket at the same voltage, so that a spark can't build up between them.

To stop the rocket melting

Inside the combustion chamber, the gas reaches around 3,300 Kelvin.

Only a tiny fraction of the internal energy in the gas flows out into the walls, but even this tiny fraction is a large amount of heat; more than enough to melt even high-melting temperature metals. So whether the rocket is made of metal or carbon fibre, it has to be protected from melting.

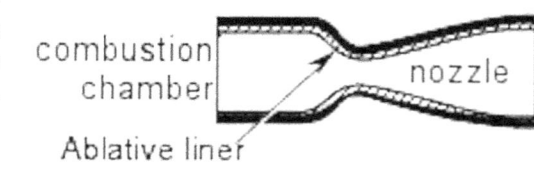

With solid or hybrid combustion chambers, a heatproof liner is fixed into the inside of the combustion chamber and nozzle:

This liner is usually made of either paper, cloth, or composite fabric, because they are very effective heat insulators.

The liner is soaked in a **phenolic resin** (which sets hard) to stop it burning. Phenolic doesn't burn, it just chars very slowly like charcoal. When it gets hot, it vaporizes into a gas that effectively forms an insulating gas barrier, and this flow of gas carries the heat away and out the nozzle.

It's important to make the liner thick enough that it hasn't eroded away completely before burnout, but not too thick that it's heavy. The slow charring and gasification of the liner is termed 'ablating', so we term the liner an **ablative liner**.

Chamber liner

We worked out the chamber liner inner diameter of 133 millimetres in the previous chapter; how thick must the liner be?

'Humble' (see references) quotes a regression rate of glassfibre/phenolic as 1.5 millimetres per second (this is a worst-case result for a position near the hotter throat).

So with our burn-time of 55 seconds, that gives a liner of:

$1.5 \times 55 = 83$ **millimetres** thick,   equ. 10. 9

worst-case.

So the combustion chambers have an inner wall diameter of:

$133 + 2 \times 83 = 299$ **millimetres**   equ. 10. 10

The nozzle

The entrant and exit sections of the nozzle would be protected by a better (more resistant) grade of phenolic composite such as silica-phenolic.

Throat protection

The throat of a rocket nozzle suffers the worst heating, because all the hot gas within the chamber is trying to flow through a small opening (giving a maximum mass flux). The throat gets protected by a mineral called **graphite** (also used to make the 'lead' in pencils) which has a very high melting temperature.

Here's a photo of a small graphite nozzle throat that I made, and also a little ring of cloth soaked in phenolic. Graphite works well as a throat protector, but graphite is brittle; it can crack, especially if made into too thin a section, or if sharp-edged grooves are cut in it.

Instead of graphite, you can use reinforced **carbon-carbon**. Carbon-carbon is cloth and resin glue like the phenolic chamber liner. Once formed, it is then pyrolyzed: heated in the absence of oxygen so that the resin glue (the matrix) decomposes and the volatile parts evaporate to leave a pure carbon matrix. Since the matrix is now very porous, a gas containing carbon - such as acetylene - has to be passed through the part so the pores will be filled with carbon.

Carbon-carbon will resist very high temperatures, and the thermal shock of suddenly being hit by hot gas at engine ignition. Because, in a sense, the material has already been burnt once, it can't be burnt again in the nozzle.

To stop the rocket leaking

We only want the gasses to flow out of the nozzle. But combustion chambers are usually made out of several sections joined together.

They're usually a central tube with an end cap at one end, and the nozzle holder at the other:

The joints between the sections mustn't leak despite the very high pressure inside.

One way to seal the joints is to weld them, or seal them with glue, but if you do that, you'll never be able to get the sections apart again. You might want to take the end off the chamber to load more fuel in for another firing, or to put the igniter in.

One way to seal the chamber in such a way that you can take it apart again is to bolt the sections together and use **O-rings** at the joints.

O-rings are just rubber rings, but they perform extraordinarily well. When you put the O-ring into the groove you've cut for it, it just sits there. (This diagram shows a section cut through the end-cap and the O-ring):

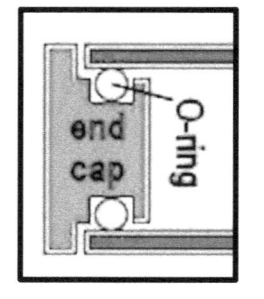

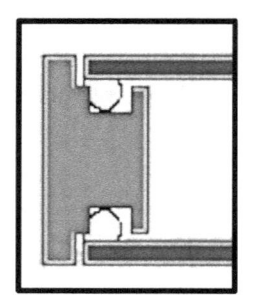

But when the chamber is pressurised, the pressure pushes on the O-ring. It gets squashed, and flows into the gap, sealing the joint:

O-rings work amazingly well for something so simple. They rarely leak, and will even flow into - and seal - nicks or scratches in the metal.

You have to choose the right rubber for rocket combustion chambers, one that doesn't melt too easily. (But you still need to protect the O-ring from the heat.) We tend to use nitrile or Buna O-rings (see chapter 8).

You also have to make sure when you fit the pieces of the chamber together that no sharp edges or sharp-edged bolt-holes nick or tear the O-rings.

Finally, you mustn't pick a rubber that gets hard when it gets cold or the O-ring won't be able to flow into the joint. The Space Shuttle Challenger was destroyed when an O-ring got too cold and failed to seal, letting hot combustion chamber gas flood out and blow up the liquid propellant tank.

Consider O-rings for the ablative liner that protects the combustion chamber wall: when the chamber of even an HPR-sized hybrid pressurizes it will grow a millimetre or two in length, which can open up gaps at either end of the liner that need to be sealed.

It's most important to use the correct size for the grooves that seat the O-ring. The correct dimensions for the grooves can be found on the web, such as O-ring manufacturer's websites.

Tolerance

Tolerance is the measure of how accurately you need to build an engine component: how many decimal places of precision are required by each component.

A liquid-propellant injector, for example, requires very high tolerances so it doesn't leak, and so that it passes the correct mass flow rates of propellants.

Hybrids are much more forgiving, which is why amateur groups prefer them. Low tolerances will suffice: perhaps 0.1 millimetres.

We at Aspirespace design our HPR-sized engines with large O-rings to mop up any gaps between components, and to deal with the fact that seamless aluminium tubing (that forms the tanks and combustion chambers) isn't always as round as it ought to be; it can be somewhat oval in cross-section.

The phenolic liner tube that protects the wall of the combustion chamber should be 'snug' (a push-fit) against the chamber wall to eliminate an air-gap there.

Chapter 11: Testing a hybrid rocket engine safely

In the previous chapters we designed a large hybrid engine large enough to get you into Space. In this chapter, it's time to discuss testing it.

If you haven't joined a rocketry society such as UKRA yet (www.ukra.org.uk) now is the time to do it.

Rocket testing can be dangerous especially if done in an unprepared manner, and you'll need special testing facilities. Rocketry groups can advise you how to keep safe.

Keep safe

Rockets generally blow up the first few times they're test-fired until you get everything right, even hybrids. When they do blow up they can maim or kill you or any others who happen to be around. So *always* assume a rocket engine will blow up.

Usually, the combustion chamber wall is metal; when it blows up, this turns into a myriad of razor-sharp red-hot fragments of metal flying through the air at several hundred kilometres an hour. If these hit you, they'll chop you to pieces. (Even if the chamber wall is made of plastic, plastic shards can still kill you if it goes bang.)

So when they're testing rockets, rocketeers hide behind thick brick walls (one thickness of bricks isn't enough to withstand the blast even from an HPR-sized explosion: obviously, the bigger the engine, the thicker the wall has to be.)

Rocketeers also need a specially built building with a thick roof, as large bits of blown-up rocket might come falling out of the sky. Any windows need glass many centimetres thick so as not to shatter. Rocketeers might instead use a periscope to peer over the wall to protect their eyes.

A building that is tough enough to withstand a rocket explosion is generally called a **blockhouse**.

Here's one of the testing sites Aspirespace has used with the site's blockhouse in the background; the walls are nearly a metre thick.

That's me on the left in the foreground looking down at our latest rocket. It's fired by remote control from the blockhouse once we're all inside.

The tank pressure, chamber pressure, and the thrust are measured by electrical sensors which are connected to wires leading into a computer in the blockhouse.

Make absolutely sure nobody else can wander into the danger area while you're testing.

The test stand

To test your rocket you want to ensure that your rocket is firmly strapped down to the ground, you don't want it to break free.

You need to bolt it to a **test stand** (or test cell).

Here's Aspirespace's test stand for our H20 liquid oxygen hybrid:

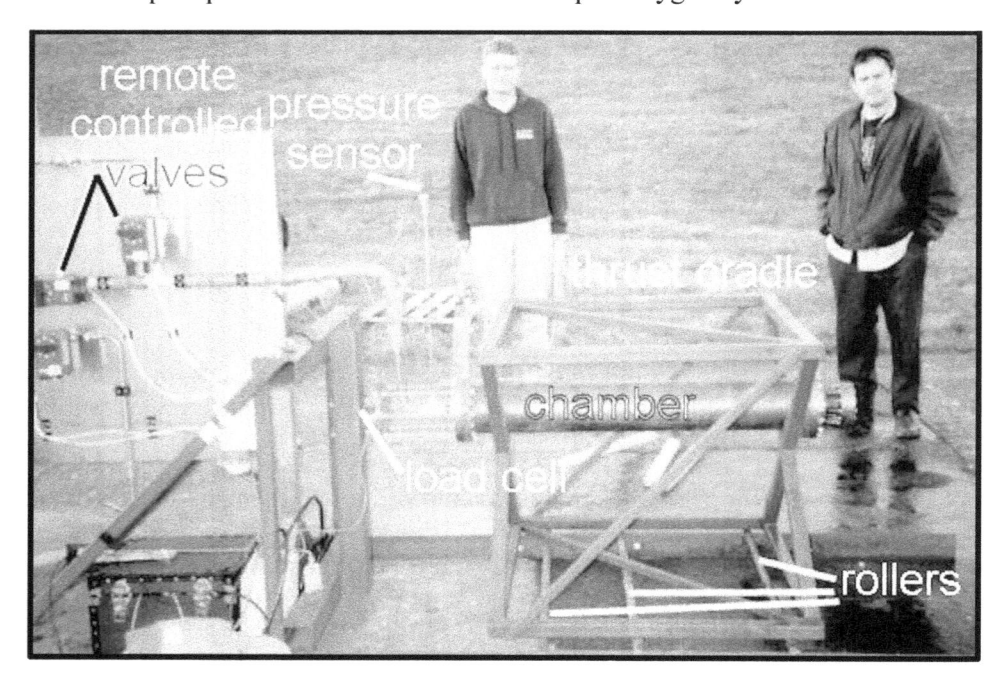

The rollers let the rocket move a little to butt up against the load cell, which is an electrical device that converts force into an electrical signal. This measures the thrust, and sends it down an electrical cable to be read by a computer.

Not shown in the picture above are heavy straps that hold the chamber and thrust cradle down to the ground, but let it move along the rollers a bit.

Right way up

When we test our Nitrous hybrid, it'll be important to remember to mount the Nitrous run-tank so that it's upright, as it will be in flight. The outlet of the tank must be at the bottom so that liquid Nitrous comes out first and not vapour.

Assembling the plumbing

When you screw the engine's plumbing together, you'll often wrap plumber's tape (PTFE tape) around the screw-threads of the fittings at the joints. Be very careful not to let strands of PTFE come loose and fall into the pipework as they can block injector orifices or jam valves.

Be careful not to over-tighten stainless steel fittings as they can 'pick-up' (cold-weld) i.e. jam solid.

Wear safety glasses and welding gloves if you need to undo any fittings on plumbing containing Nitrous. Wear long-sleeved clothing, and avoid man-made fibres as they can melt in a fire and scald your skin with red-hot melt.

Earthing

It's also important to **earth** the hybrid. This means connecting all the metal parts of the hybrid together with wires, and then connecting these wires to a metal pole stuck into the ground (into the Earth). This allows any static electricity that builds up to get channeled away into the ground. If the test-stand is metal, earth the hybrid to the stand and then earth the stand into the ground.

Make sure to electrically connect your run-tank to the fill-tank before filling in case they're at different voltages: you don't want a spark in the fill pipe setting off the Nitrous there, or *both* tanks could blow up.

Measuring pressure

Record combustion chamber pressure if you possibly can as it allows you to work out C* (see page 66).

To measure the tank and combustion chamber pressures you need to make a tiny hole in the tank and in the chamber. These holes (called **pressure taps**) are connected to very narrow diameter pipes that lead to pressure sensors. Using only a small hole means that the flow won't be disturbed unduly by the hole.

You'll need to make the narrow pipes about 30 centimetres long and running upwards with the sensor on top. Any longer and the gas trapped in the pipes won't be able to react quickly enough to sudden changes in flow pressure, but any shorter and the hot combustion chamber gas could get all the way up the pipe and melt the sensor. The sensors convert pressure into an electrical signal.

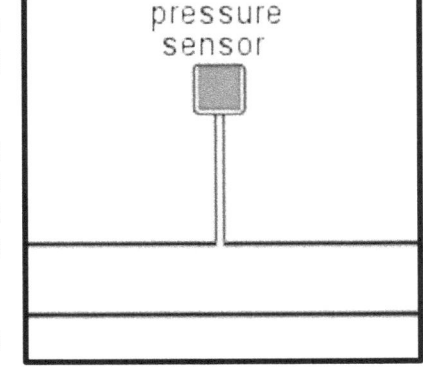

Interestingly, because the mouth of the pressure tap is at 90 degrees to the direction of the gas flow, it measures 'static' pressure (see page 15).

Fortunately, within the tank and combustion chamber, the 'static' pressure and stagnation pressure are pretty-much the same because the flow is moving so slowly.

Data logging

Getting the signals from your pressure and thrust sensors into your computer is called **data logging** and is a whole book in itself on electronics.

Here are some brief hints and tips from an ex-Aspirespace electronics expert:

- Avoid earth-loops (called ground-loops in America). See wikipedia: Ground-loop (electricity)

- Try to use transducers drawing between 4 to 20mA of current where possible.

- Try to buy sensors which are factory pre-calibrated to a known accuracy: it saves a lot of calibration work.

- Strain-relieve all wiring: make knots or connections within the plugs connecting your wiring so that the wires can't be pulled out of their plugs if anyone trips over the wires.

- Mount pressure sensors so that they don't get suddenly hit with cold liquid on the sensor diaphragm (e.g. keep the pressure-tap pipe upright).

- Don't let the data-logging electronics accidentally fire the engine igniter: galvanically isolate (electrically isolate) the data recording system from the final link/safe-arm circuit and from any pyrotechnic firing circuits using opto-isolators, relays, or hall-effect current sensors as applicable.

- Electrically screen the thrust-measuring load cell cables, but connect the screen at the computer end not the load cell end.

- Have one single earth lead from the test stand to the computer.
- Ensure you have lots of spare batteries and that they are all fully charged.
- Test the whole electrical system before going to the test site.
- Use appropriate connectors and cabling. Try not to use mains connectors on low-voltage systems in case you accidentally plug them into a high voltage supply.

Test-site safety

Before you fire the engine, think very hard about test-site safety, which is similar to rocketry range-safety as discussed in the next chapter.

Ensure you know where all people will be during the test. Ensure there are clear lines of communication: you don't want anyone not to hear that a firing is just about to happen. Designate one person as the Safety officer, who has responsibility to ensure that the site and people are safe before and just after firing, and must be obeyed.

Immediately after firing, don't all jump up and down cheering. Make sure everything is safe first, *then* jump up and down.

Static-firing misfortunes

The following - hopefully educational - examples describe what might happen to your rocket engine when you test it. But they're not fiction, they're based on actual events that have happened to me or mutual friends (you have been warned!).

Example 1

Assuming that your large hybrid rocket passed its hydro-testing and leak testing satisfactorily (chapter 8) then it's time to test-fire the rocket for the very first time. When you fire a rocket tied to the ground, it is called a **static-firing**

You fire the engine's igniter by remote control from your blockhouse. But no flame comes out the nozzle, there's just a white cloud of Nitrous coming out, and the thrust is feeble. The engine didn't light, what happened?

The mass flow rate of Nitrous out of the tank and into the combustion chamber depends on the pressure difference between the tank and the chamber. You get the biggest difference in pressure just as the engine starts, when the combustion chamber is still just at the pressure of the atmosphere outside the engine (that's the lowest it can go) so the mass flow rate of Nitrous into the chamber is very large right at the start.

The mass flux down the port was too high, and blew the flame out. You must increase the initial port diameter to lower the flux. Annoyingly, the lower flux means that the regression rate will drop.

This means you have to make the fuel grain longer to harvest off enough fuel, which needs a longer combustion chamber to be built.

Just how high a mass flux you can get to light up depends mainly on how much energy you throw at it: don't hold back on throwing solid propellant at ignition. It also makes a difference where you put the heater charge. The closer it is to the injector, the better it works because then it breaks the Nitrous down into hot oxygen before the Nitrous hits the fuel. (But don't set fire to the injector.)

Example 2

Just as the engine fires into life, the flame that was coming out the nozzle vanishes…

… and then the engine blows up.

What went wrong?

The vanishing flame is the clue: something blocked the nozzle, stopping the flame coming out.

If rockets have their nozzle blocked, they generally explode. Too much gas builds up inside the combustion chamber with no way out, so the pressure inside goes sky-high, and the chamber wall bursts. You made the heater charge too wide. It happened to break loose, and was wide enough to block the nozzle throat.

Example 3

You fire the hybrid again. It seems to be working…

Bang!

Something large and hot blasts out of the end of the hybrid, flies a long way through the air at very high speed, then buries itself deeply into the ground.

Your nozzle chunked: a big chunk of the graphite ring at the throat got spat out. Graphite's quite brittle, you've got to treat it gently. Perhaps it got dropped in your workshop and cracked.

Or perhaps it got damp. Graphite will happily soak up water like a sponge. When the rocket's firing, this water flashes into high-pressure steam that expands greatly and so can split the graphite apart. It would be best to bake your graphite in an oven for an hour or two before rocket firings from now on, to get the water out.

Example 4

You fire the hybrid again. It works, but when you look at the thrust, you find that it's way lower than it ought to be, and so is the chamber pressure.

The thrust is lower because you're testing at ground-level instead of 21.3 kilometres up, so the atmosphere's back-pressure on the nozzle exit is much higher.

This drops the effective exhaust velocity down to 2,247 metres per second (I_{SP} of 229) at the start of the burn which is 5/6ths of what you'll get at altitude, so the ground-level thrust is only 5/6ths of what you will get at altitude. [The characteristic velocity C^* doesn't change with altitude but the thrust coefficient C_F *does* change with altitude, it will be 5/6ths of what you'll get at altitude. Remember that $Isp = \dfrac{C^* \, C_F}{9.81}$]

For static firing on the ground, the atmosphere's back-pressure is high, so we only need a nozzle area ratio of five.

Example 5

The hybrid's working, but towards the end of the burn it's giving off a horrible farting noise (pardon my French, but the noise is unmistakable) which is always the sign of a hybrid in trouble.

A hybrid's combustion chamber pressure is never quite constant; it wobbles up and down a few Bar several times a second. This is because the eroding of fuel off of the surface of the plastic port is rough and unsteady. Sometimes more fuel gets worn off, sometimes less; so sometimes there's more mass inside the chamber, sometimes less.

These pressure wobbles can be dangerous. When the pressure bounces up a few Bar, the chamber pressure can briefly end up higher than the run-tank pressure.

This makes the flow reverse its direction: hot combustion chamber gas flows backwards through the injector and heads towards the run-tank.

The horrible noise warns you that the flow through the injector is switching from going forwards to backwards many times a second.

Bang!

Hot combustion chamber gas got all the way back to the run-tank and set off the Nitrous.

The whole tankful of Nitrous decomposed, giving off so much heat that the Nitrous expanded twenty times and so the tank pressure went sky-high, and the tank burst like a bomb. A lot of your test-stand got destroyed in the explosion.

Why didn't your run-tank burst disc break to save the tank? You forgot to fit it, instead you left the big unbreakable disc used for the hydro-testing blocking the emergency outlet pipe. (Don't laugh, I know someone who did this.)

Your hybrid had terminal flatulence (see chapter 8): what can you do about this; you need to stop the reverse flow from happening.

The way to do it is to make sure that the chamber pressure never bounces higher than the tank pressure. This means that you've got to lower the chamber pressure, so you have to drill the nozzle throat wider to let the gas out more easily: recall chapter 4 page 52 where I discussed the balance between mass flow into and out of a combustion chamber. Sadly this means that the thrust will be lower, but safety comes first.

A Nitrous hybrid's tank and chamber pressures are closest at the end of the burn, which is why your hybrid ran into trouble there.

In chapter 9 I suggested a nozzle throat of 68 millimetres wide to get a mid-burn combustion chamber pressure of 33 Bar. This should give a chamber pressure of around 28 Bar at the end of the liquid burn, when the tank pressure is about 35 Bar. (The problem is worst at the end of the burn when tank and chamber pressures are closest.)

Then:

$$\Delta P = 35 - 28 = 7 \text{ Bar} \quad \text{equ. 11. 1}$$

that's a bit more than $1/5^{th}$ (20%) of the chamber pressure so it should be safe: a minimum of 20% is recommended at the end of the liquid burn.

Example 6

The engine's made it on one piece right until burnout…

Bang!

…and then just at burnout, the run-tank burst disc blows, and Nitrous vapour spews out the emergency dump pipe.

But it's not just gushing out, it's *roaring* out.

When you look at the footage from your digital video camera, you can see that the gas rushing out of the pipe is very hot. (Cheap digital cameras often pick up infrared 'heat' radiation as well as light, which is useful as you can spot hot-spots on your rocket.)

Somehow, the Nitrous vapour remaining in the tank just at burnout decomposed into hot gasses which expanded in their own heat and sent the run-tank pressure way too high which burst the burst disc.

The problem occurs at the end of the burn when all the liquid Nitrous has run out. This just leaves Nitrous vapour at lowish pressure in the plumbing and the run-tank. The flames from the combustion chamber have gotten backwards through the injector and set off the Nitrous vapour. The Nitrous then decomposes all the way up the plumbing and into the run-tank; the tank pressure goes sky-high and the burst disc bursts.

The way to stop the flames spreading backwards up the plumbing is to kill the decomposition.

This problem is rare because it only occurs with large Nitrous hybrids. It doesn't occur at all with the small HPR hybrids you can buy from rocketry vendors (which use only a few litres of Nitrous) for a rather interesting reason: as discussed in chapter 8, it turns out that the decomposition can't pass through small metal pipes such as the feed pipe between the tank and the injector.

There's too much metal (the pipe wall) too close to the Nitrous. The metal soaks up the heat of decomposition, killing further decomposition. This is called quenching (chapter 8). But with large Nitrous hybrids, the Nitrous vapour in the middle of a *large* feed pipe is too far away from the metal wall to leak its heat into the metal.

There are two ways that have been discovered so far to kill the decomposition, though one of them's just an idea of mine that hasn't been tested in practice yet:

Dilution

It's been found (chapter 8) that if you dilute the Nitrous vapour with helium gas, this gas soaks up the heat from the decomposition, killing it. This has been tested and it works.

It's not yet clear just how much helium you have to put into the run-tank for safety. It appears to be about 1/10th of the total mass of Nitrous at the beginning of the burn which is rather a lot of extra useless mass to carry.

Annoyingly you'll need a bigger run-tank to cope with this added helium which adds yet more mass.

[Helium's quite expensive: other gasses would be cheaper, but helium has a particularly high thermal conductivity so soaks up heat particularly well. Oxygen can be used, and has the advantage that it is a Nitrous-compatible oxidiser.]

Quenching

We know that the decomposition can't pass through small metal pipes (less than 6 millimetre internal radius according to tests that have been done with Nitrous, see chapter 8), such as the feed pipe between the tank and the injector on small hybrids. This is due to quenching.

Of course the question then is: how did the decomposition get backwards through the metal injector without quenching?

I think it has to do with the fact that the injector orifices have a short length, and so there isn't enough surface area of metal inside each orifice to soak up the heat (heat transfer depends on the surface area you're trying to transfer the heat through). Also, a chamber pressure-pulse could squirt the decomposed Nitrous vapour back through the injector swiftly, not giving enough time for the heat to flow out of the Nitrous and into the injector orifice.

So I think that a simple cure for large Nitrous hybrids would be to put a bundle of lots of little metal pipes lengthwise into the big feed pipe just upstream of the injector. Provided that each pipe was long enough then it would have enough surface area to soak up the heat of any decomposition.

A bundle of 6 millimetre diameter - or smaller - aluminium pipes will do. (Each pipe is parallel to the flow direction.)

Wire gauzes are often placed across pipes to prevent flames going upstream past the gauze. These are called flame traps, so you could call my idea of a pile of metal pipes a 'quench trap'.

But I haven't tested this idea yet.

Example 7
You fire the hybrid up, it seems to be working! There's no nasty noise anymore, but just as you're feeling confident…

Phut!

A flame appears coming out sideways from the end of the combustion chamber just ahead of the nozzle.

You've had a burn-through. Your phenolic insulation in the post combustion chamber wasn't thick enough and it all burnt away, leaving the bare metal of the combustion chamber exposed to the scorching hot exhaust gas. This metal melted and this made a hole.

Perhaps you unwisely made a joint in the insulation: when the combustion chamber pressurised, the chamber grew in length a few millimetres (strain, see page 133) and this joint opened up into a gap. Seal the gap with an O-ring next time.

On with the rocketry

When you eventually achieve success, you can then build carbon fibre combustion chambers to get the weight off.

To have a rocket that is safe enough to trust, you'll want *at least* twenty completely successful static firings in a row before you even *think* about sitting on top of it. Some commercial rocket engines have been test-fired 200 times.

Chapter 12: Flying a hybrid

This chapter is a brief guide to the legality and practice of flying hybrids. We've looked at static testing hybrids in the previous chapter, but actually flying a hybrid presents some unique challenges.

The main issues are listed in the following paragraphs.

Air law

Before flight, be sure you have acquainted yourself with local and national aviation laws governing the flight of amateur rockets. The onus is on you to find out the laws that apply in your particular country. Generally, the higher you intend to fly, the more onerous the paperwork.

It's prudent to join - or at the very least contact - your national rocketry society in order to learn the national laws, and most importantly, how to construct the vehicle around the rocket engine in a manner that will ensure its structural integrity and recovery.

In Britain, for HPR rockets, download the latest United Kingdom Rocketry Association safety code at 'safety code' at www.ukra.org.uk

In other countries, it's your responsibility to follow that country's own rocketry safety code (such as NAR or Tripoli in the USA).

The ethos of air law is to protect the public. It cares less about you, the rocketeer, but is very stringent about protecting uninvolved persons in the air and on the ground.

Range safety

The safety of uninvolved persons on the ground under the flight of your rocket vehicle is known as **range safety** (from the phrase 'launch-range'). Again, rocketry societies can advise on range safety.

You need a circular area clear of people or buildings, and the higher you're going, the wider this circle has to be. That's a very wide area if you're going to Space.

You must analyse trajectory simulations and consider failures so as to predict all directions your vehicle could end up flying, and how far it could travel horizontally in the event of a launch mishap. If your vehicle does go off-course, have a system available that can shut off the engine and drain the oxidiser tank before it lands.

One of the benefits of hybrid systems is that they very rarely explode in flight. This reduces the likelihood of exploded fragments travelling a distance, but for safety analyses assume that it could happen.

Flight weight

On the ground, the weight of the hybrid system is unimportant. Indeed thick, heavy, tank and combustion chamber walls are safer as they can withstand higher stress.

But from the rocket equation (equation 9.9 chapter 9) it's vital that we keep the empty mass of the rocket vehicle as low as possible, which means keeping the mass of the propulsion system as low as possible as it is usually the largest part of the empty mass. The same applies to the remaining mass of the vehicle structure.

Growth factor

The key to an efficient rocket vehicle structure is keeping the mass down. The effect of the need to keep the structure mass low can be neatly described by the 'growth factor' which is proportional to the mass ratio (see equation 9.10) required for the mission.

For a car, every extra kilo of structure will require perhaps a 1/6th kilo of extra fuel, which requires a bigger fuel tank, which requires more structure, which requires more fuel…round and round this mass-growth loop goes until it settles down at a heavier car. The growth factor is the number that indicates that for every extra kilo of structure you add to the car, another X kilos of structure and fuel will appear to carry it.

The growth factor of a car is low (less than one). For an aeroplane, the growth factor is around four, so now weight control is a real issue if the takeoff mass isn't to spiral out of control.

But for vertically-ascending spacecraft launched from the ground, the growth factor is around ten, so weight control is absolutely critical: if the mass gets out of control, the vehicle might get too heavy to even take off.

Fortunately for suborbital spacecraft launched from a high altitude gas balloon, the growth factor comes down to more manageable values because the mass ratio required is much lower.

Aim to make structural components do more than one job simultaneously. But balance that against the need for **structural redundancy** which means that if one part breaks, some other part can pick up the load.

The engine

We must keep the weight of the propulsion system: tank, plumbing, and engine, as low as possible.

The tank and combustion chamber should have adequate - but not excessive - safety-factors to keep their mass down. Pay particular attention to the weight of combustion chamber end closures as these can get heavy if poorly designed: a little extra machining can remove mass that isn't doing anything structurally useful.

On smaller HPR hybrids, valve actuation systems can get heavy: a piston actuator powered pyrotechnically by expulsion powder (see page 95) has a much higher power-to-weight ratio than pneumatic or electric actuators which require the mass of stored gas bottles or batteries.

[A very small choke/orifice between expulsion powder and piston actuator can provide a slowly-opening valve if required.]

Does the run-valve (see page 37) need its actuator mounted within the rocket vehicle, or could it be bolted to the launchpad via a shaft that disconnects at launch?

Commercial valves themselves can be heavy as they're not designed with weight in mind. It's often possible to modify or re-design valves to reduce mass, but be sure to test that they will work reliably and repeatedly, and will withstand the fluid pressure with an adequate safety-factor.

Hybrid ignition systems can get heavy when carried on the vehicle. Unless it's purely pyrotechnic (e.g. a pyrovalve system, see page 94) it's better to mount the ignition system to the launchpad acting up through the nozzle throat.

Remote operation - the launch system

For the safety of launchsite personnel, the launch has to be conducted by remote control from a firing position at some distance from the launchpad. National rocketry safety codes (e.g. UKRA) will stipulate mandatory safe distances between the launch personnel and the launchpad: this distance increases with the size and power of the engine.

As discussed in chapter 3, filling the run-tank remotely allows a smaller safety-factor on the run-tank: a lighter tank.

Pre-launch and aborts

Remote-fill systems must be designed to fail-safe, i.e. will empty the run-tank if electrical or pneumatic/hydraulic power is lost, otherwise there's no way to depressurise a remote-filled run-tank if the remote-fill system dies and the hybrid didn't launch. You can't approach the launchpad without serious risk to yourself unless the run-tank has a large safety-factor.

The simplest approach for Nitrous hybrids is the one commonly used; the vent-hole's secondary function is a deliberate leak to slowly lower the tank pressure in the event of an abort. A deliberate leak is a good idea even on a vent-free system.

For safety, vent all remaining Nitrous in an aborted fill, and wait at least 15 minutes before re-filling otherwise the tank will be excessively cold and thrust will subsequently be lost.

Launch

Most rocket engines are ignited electrically in some manner, as remote operation is easy to implement electrically. Generally for HPR vehicles, the battery to power the ignition system and any valve actuation required for launch is situated at the launchpad, with a low-power remote cable extending out to the launch operator's position. The low-power cable actuates mechanical or solid-state relays connected to the launchpad battery to command the launch electrical power.

If used, a remote-filling mechanism absolutely must disconnect from the rocket vehicle at or before launch, otherwise the majority of the launchpad gets carried aloft and the rocket vehicle crashes. Following the advice of NASA literature on remote disconnection systems, consider backup mechanisms in case the primary disconnection mechanism jams.

The Reaction Research Society has launched rockets using an ingenious re-rig of a standard hydraulic quick-disconnect coupling: The coupling's release-collar is tied via a lanyard to a stake in the ground, and the male part is fixed into the rocket pointing aft. When the rocket pulls upward on the coupling at launch, the lanyard tied to the release-collar goes taut and the coupling disconnects automatically. But note that many commercial quick-release couplings can't be easily disconnected before first de-pressurising the coupling.

Alignment

Many traditional rocket vehicles are axisymmetric: they exhibit symmetry around their central axis, and are built up from tubular components.

Whether axisymmetric or not, it's vital that the thrust line of the engine passes through the centre of mass (**centre of gravity**: C.G.) of the vehicle to high precision. As rocket thrust is such a large force, even a tiny misalignment can cause a large torque around the C.G. which can overwhelm the stability system of the vehicle. The thrust line depends on the critical alignment of the nozzle axis of symmetry, and the structural integrity of the nozzle during firing.

Similarly, the C.G. of the propulsion system itself must be aligned so as not to pull the overall C.G. of the vehicle off to one side away from the thrust line. Hybrid propulsion system components (tanks and combustion chambers) tend to be physically long, so are easily angularly misaligned off the vehicle axis. Consider a flexible hose between tank and combustion chamber/s to ensure alignment of each separately.

For our manned Space vehicle, one particularly large mass that has to be aligned carefully is you the occupant!

Aerodynamics

First off, don't forget the aerodynamic alignment. The overall centroid of aerodynamic forces on the vehicle (the **centre of pressure**) also has to be laterally aligned with the vehicle C.G. and thrust line to prevent rotation. This is achieved by accurate fins and nosecone alignment.

Aerodynamic drag can be one of the largest forces suffered by a rocket vehicle. If launched from the ground, the drag force is only slightly lower than the thrust, though for our vehicle launched from a gas balloon it's significantly lower.

The magnitude of the drag force (so in effect the influence of the launch altitude) determines the maximum allowable diameter of the vehicle: if launched from the ground, air density is high so fuselage diameter (cross-sectional area 'S' see equation 1.34, chapter 1) has to be minimised to reduce drag as much as possible. This results in a long, thin propellant tank for a hybrid which weighs more: the optimum length-to-diameter ratio of the rocket vehicle which optimises vehicle mass against Drag loss (see equation 1.36) is around 20:1 for small HPR vehicles, reducing with increasing vehicle size.

Conversely, our rocket vehicle launched from high altitude can be fat and compact with a minimum-mass spherical propellant tank as sketched on page 125.

The effect of Drag loss on an HPR-sized vehicle launched from sea-level is dramatic. For a high acceleration commercial solid-propelled rocket vehicle, its apogee would be roughly nine times higher if launched in a vacuum. An HPR-sized hybrid-powered rocket vehicle can operate at lower thrust, to reduce flight velocity to reduce drag and increase apogee. It might experience a height increase of four times in a vacuum, but note that its apogee would be significantly higher than a commercial solid-propelled rocket vehicle of the same diameter.

[Commercial HPR hybrids generally won't outperform a solid motor because their tanks and combustion chambers are designed with extremely large safety-factors because of the worry of customer litigation and damage due to parachute failure. Therefore they are considerably over-heavy.]

As we saw from equation 5.13 chapter 5, the effective exhaust velocity and specific impulse increase at altitude, because the reduced air pressure increases the nozzle thrust coefficient. This effect is dramatic, causing a large engine performance increase that greatly increases the apogee of a rocket launched at altitude.

Including both the drag decrease and thrust coefficient increase, the Aspirespace 'Rickrock' hybrid vehicle if launched from sea-level will achieve a height increase above launch altitude of 5.5 kilometres. If launched at 21.3 kilometres, it will achieve a height increase above launch altitude of 25.4 kilometres.

A suitably designed HPR-sized hybrid (of around 'M' class total impulse) could have a high enough mass ratio to reach Space if launched at around 100,000 feet (30.5 kilometres).

Trajectory

HPR

HPR-sized hybrid-powered rocket vehicles are generally passively stabilised by aft-mounted fins. These fins are ineffective below around 20 metres per second airspeed, so it's customary to use a launch-rail to stabilise the vehicle for the first second or so of ascent until this airspeed is reached.

Due to physical limitations on their fuel grain regression rate, hybrids provide a lower thrust than commercial HPR solid motors, so a *considerably* taller launch-rail is needed to attain this airspeed.

Fin-stabilised rocket vehicles essentially follow a **gravity turn trajectory** wherein gravity magnifies any non-verticality of the trajectory and pulls the trajectory over to the horizontal and then downwards.

[The larger the vehicle's aerodynamic static stability, the more closely it approximates a gravity turn.]

The equation for a gravity turn trajectory is:

$$\frac{d\theta}{dt} = \left(\frac{V}{r} - \frac{g}{V}\right)\cos(\theta) \quad \text{equ. 12.1}$$

where θ is the trajectory angle of the vehicle relative to the local horizontal, and so $\frac{d\theta}{dt}$ is the pitch angular (rotational) velocity. V is vehicle velocity along the flightpath, g is gravity, and r is the distance from the vehicle to the Earth's centre.

Even for our suborbital mission to 100 Kilometres, assuming that the Earth is flat incurs minimal error, so the term involving r drops out ('infinite' r) to give:

$$\frac{d\theta}{dt} = -\frac{g}{V}\cos(\theta) \quad \text{equ. 12.2}$$

This equation tells us that the speed that the trajectory bends over is inversely proportional to the vehicle's speed V, therefore HPR solid motors are provided with a 'boost thrust' just after ignition to get the speed that the vehicle exits the top of the launch-rail as high as possible to preserve verticality for as long as possible.

[The speed that the trajectory bends over is also proportional to how far over the trajectory already is (the cos (θ) term).]

HPR hybrid fliers have discovered that this lack of initial boost thrust causes a trajectory that rapidly arcs over, sending the vehicle on a low, flat, trajectory over the horizon. The large horizontal airspeed at apogee then overtaxes the parachute recovery system which subsequently fails.

Note that even the slightest wind will cause the trajectory to bend off the vertical as it leaves the launch-rail because the wind causes an **angle of attack** on the nosecone and fins that rotates the vehicle about its C.G.

Thus it is necessary to provide fin-stabilised hybrid rocket vehicles with:

1. The tallest launch-rail possible to allow acceleration to reasonable rail-exit velocity.

2. A boost thrust. The boost thrust is provided by strap-on boosters or a lower stage. This booster/s engine can be powered by solid propellant (commercial HPR motors) or compressed gas (steam, air, air-plus-water) but only needs to have a burn time of a fraction of a second so the booster propellant requirement is low. Shown here is Jonathan Rhode's 'Spirit of Columbia' M-class hybrid rocket vehicle with twin strap-on solid boosters:

Larger vehicles

Our manned spacecraft will require an active control system to maintain its verticality. This would be a very simple autopilot rather than the unnecessary complexity of a guidance system: simply three orthogonal rate gyros linked to a control computer, as used on drones and smartphones.

The autopilot would be linked to a device that angled the thrust line (known as **thrust vectoring**).

Nitrous hybrids have a relatively low exhaust gas temperature at the nozzle exit, which allows simple mechanical vanes/rudders and tabs to be inserted into the exhaust to steer it. One promising device Aspirespace are investigating is the **jetavator** which is a metal ring that encircles the nozzle exit, and is tilted on gimbals to deflect the exhaust:

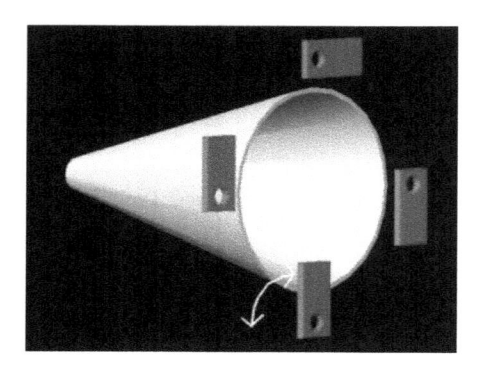

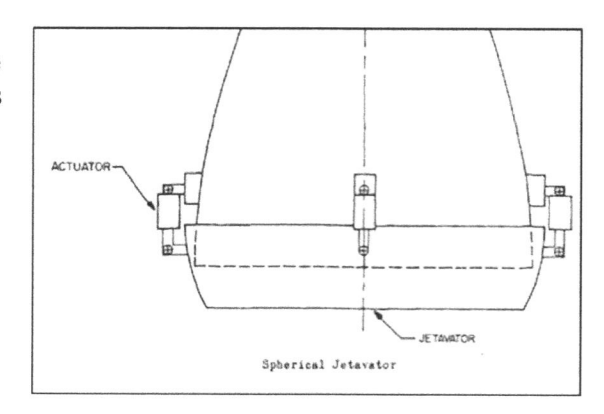

Spherical Jetavator

Appendix 1: Earth's Atmosphere

The International Standard Atmosphere is a model of Earth's atmosphere from sea-level to 80 kilometres altitude, it's listed below.

I've used data from Engineering Science Data Unit (ESDU) 77021 for 0 to 80 kilometres, and above (many thanks to ESDU for permission to reproduce the following).

'E' is Microsoft Excel for '10 to the power of'.

Above 90 kilometres, the speed of sound loses its meaning because the atoms of air are too far apart to allow supersonic effects such as shockwaves, so I've assigned an arbitrary large number to keep the Mach number small (Mach number = velocity divided by speed of sound.)

Geometric Height above sea level (kilometres)	Temperature (degrees C) (K)		Pressure (Pascals)	Natural log of Pressure	Density (kilograms per cubic metre)	Natural log of Density	Speed of sound (metres per second)
0	15.0	288.2	1.01325E+05	11.5261	1.225E+00	0.2029	340.3
0.5	11.8	284.9	9.54610E+04	11.4665	1.167E+00	0.1547	338.4
1	8.5	281.7	8.98760E+04	11.4062	1.112E+00	0.1059	336.4
2	2.0	275.2	7.95010E+04	11.2835	1.007E+00	0.0066	332.5
3	-4.5	268.7	7.01210E+04	11.1580	9.093E-01	-0.0951	328.6
4	-11.0	262.2	6.16600E+04	11.0294	8.194E-01	-0.1992	324.6
5	-17.5	255.7	5.40480E+04	10.8976	7.364E-01	-0.3059	320.6
7	-30.5	242.7	4.11050E+04	10.6239	5.900E-01	-0.5276	312.3
10	-49.9	223.3	2.65000E+04	10.1849	4.135E-01	-0.8831	299.5
15	-56.5	216.7	1.21120E+04	9.4020	1.948E-01	-1.6360	295.1
20	-56.5	216.7	5.52900E+03	8.6178	8.891E-02	-2.4201	295.1
21.3	-55.3	217.9	4.51300E+03	8.4147	7.216E-02	-2.6289	295.9
25	-51.6	221.6	2.54900E+03	7.8435	4.008E-02	-3.2168	298.4
30	-46.6	226.5	1.19700E+03	7.0876	1.841E-02	-3.9949	301.7
35	-36.6	236.5	5.74600E+02	6.3537	8.463E-03	-4.7721	308.3
40	-22.8	250.4	2.87100E+02	5.6598	3.996E-03	-5.5225	317.2
45	-9.0	264.2	1.49100E+02	5.0046	1.966E-03	-6.2318	325.8
50	-2.5	270.7	7.97800E+01	4.3793	1.027E-03	-6.8811	329.8
60	-26.1	247.0	2.19600E+01	3.0892	3.097E-04	-8.0799	315.1
70	-53.6	219.6	5.22100E+00	1.6527	8.282E-05	-9.3988	297.1
80	-74.5	198.7	1.05300E+00	0.0516	1.846E-05	-10.8999	282.5
90	-86.3	186.9	1.83600E-01	-1.6950	3.416E-06	-12.5870	1000000
100	-78.1	195.1	3.20000E-02	-3.4420	5.607E-07	-14.3941	1000000
120	86.85	360.0	2.53900E-03	-5.9760	2.222E-08	-17.6223	1000000
150	361.3	634.4	4.54200E-04	-7.6970	2.076E-09	-19.9928	1000000
200	578.9	852.1	8.54200E-05	-9.3679	2.571E-10	-22.0816	1000000
300	976.0	702.9	8.77230E-06	-11.6439	1.92E-11	-24.6781	1000000
500	999.2	726.1	3.02280E-07	-15.0119	5.22E-13	-28.2821	1000000
1000	1000.0	726.9	7.51092E-09	-18.7069	3.56E-15	-33.2691	1000000

Appendix 2: Design values for a Nitrous oxide and high-density polyethylene hybrid engine

As the Nitrous oxide empties from its tank and into the hybrid, its pressure drops continuously.

The tank pressure (the vapour pressure) at liquid burnout ends up at about 2/3rds the vapour pressure it started at.

This has the knock-on effect of causing the combustion chamber pressure to reduce continuously throughout the burn, and so the nozzle mass flow rate decreases too. The nozzle mass flow rate drops to about 7/10ths (= 0.7) times the value it started at, at liquid burnout.

So if the numbers all change during the burn, which set of numbers do you choose to design the hybrid?

Hybrid rocketeers have found that a good design point is to take the tank and chamber pressures that occur midway through the liquid phase burn: the mid-burn values. These are listed here, using numbers generated by the PROPEP 3 software (see chapter 6) assuming frozen flow (so that the values doesn't change down the nozzle) for Nitrous oxide liquid and HDPE plastic with sea-level atmospheric conditions outside the nozzle.

The pressures listed below are given in absolute values (abs) which are the real pressures, rather than gauge pressures. [Mechanical pressure gauges have zero set at the atmospheric pressure around them, so they show a reading that is one atmosphere (1.013 Bar) less than the actual abs pressure.]

Start values

Initial tank pressure: 51 Bar abs

Initial Nitrous temperature: 20 degrees C

Initial Nitrous liquid density: 787 kilograms per cubic metre

Initial chamber pressure: 37 Bar abs

End values (when the liquid Nitrous runs out)

Final tank pressure: 35 Bar abs

Final Nitrous temperature: 4.5 degrees C

Final Nitrous liquid density: 884 kilograms per cubic metre

Final chamber pressure: 28 Bar abs

Mid-burn values

Tank pressure: 43 Bar abs

Nitrous temperature: 13 degrees C

Nitrous liquid density: 835 kilograms per cubic metre

Chamber pressure: 33 Bar abs

With these mid-burn pressures and Nitrous density, each square millimetre of injector orifices total cross-sectional area will let through 28.9 grams per second (= 0.0289 kilograms per second) of liquid Nitrous mass flow rate (see appendix 3).

Nitrous/HDPE mid-burn thermochemistry

oxidiser-to-fuel (mixture) ratio (by mass)	6.4:1	7:1	7.3:1	
Combustion chamber (stagnation) temperature $T_{chamber}$	3229	3281	3296	Kelvin
Isentropic exponent (ratio of specific heats) γ	1.261	1.257	1.256	
Specific gas constant R	323.4	322.7	319.4	joules per kilogram Kelvin
Molecular 'weight' (relative molecular mass) W	25.71	25.76	26.03	
Specific heat capacity at constant pressure C_p	1564	1577	1567	joules per kilogram Kelvin
Characteristic velocity C^*	1560	1557	1553	metres per second
Throat area to pass one kilo per second of nozzle mass flow rate	472	471	470	square millimeters per kilo per second
Ideal expansion sea-level nozzle (mid-burn values):				
Area ratio ϵ	4.70	4.72	4.73	perfect nozzle, no pressure thrust
Ideal thrust coefficient C_F	1.479	1.480	1.480	
Ideal (isentropic flow) effective exhaust velocity C_e	2307	2303	2298	metres per second
Ideal (isentropic flow) specific impulse Isp	235.2	234.8	234.3	seconds

For actual values, multiply C_F, C_e, and Isp by 0.98 to take energy losses into account (see page 85).

Appendix 3: Mathematical equations used in rocketry

Here listed are some equations that I've mentioned earlier in the book. (It's worth reviewing the section 'Indice algebra' in chapter 1 for the derivations below.) And remember to use Kelvin in these equations, *not* degrees C.

Young's modulus E

$$E = \frac{\sigma}{\varepsilon} \quad \text{App. 3. 1}$$

where σ is the stress applied to some material, and ε is the strain this causes which is the increase in length of the material divided by its original length.

Specific gas constant R

$$R = \frac{R_0}{W} \quad \text{App. 3. 2}$$

where W is the molecular 'weight' (also called the molecular mass; nowadays called the relative molecular mass) of the gas and R_0 is the universal gas constant (which has the same value for all gasses) = 8,314 Joules per kilogram-mole Kelvin.

For example, the molecular 'weight' of air is 28.96 therefore:

$$R \text{ for air} = \frac{8{,}314}{28.96} = 287.1 \text{ Joules per kilogram Kelvin} \quad \text{App. 3. 3}$$

Density

Density ρ is linked to mass m and volume V as:

$$\rho = \frac{m}{V} \quad \text{App. 3. 4}$$

rearranging:

$$V = \frac{m}{\rho} \quad \text{App. 3. 5}$$

and:

$$m = \rho V \quad \text{App. 3. 6}$$

Speed of sound a

The speed of sound:

$$a = \sqrt{\gamma R T} \quad \text{App. 3. 7}$$

where T is the gas temperature in Kelvin

For atmospheric air, $\gamma = 1.4$ and $R = 287.1$ Joules per kilogram Kelvin.

Mach number M

$$\text{Mach number} = \frac{gas\ speed}{speed\ of\ sound} \quad \text{or:} \quad M = \frac{V}{a} \quad \text{App. 3. 8}$$

Injector mass flow rate

We want to calculate the mass flow rate of liquid propellant that will flow through an injector due to a pressure ΔP across it. Directly upstream of the injector face is the pipe section known as the injector manifold.

We start with **Bernoulli's equation** (an energy equation which describes pressure as a form of potential energy and flow velocity as kinetic energy) for the flow of liquid from the injector manifold into the injector orifices:

$$P_{manifold} + \tfrac{1}{2}\rho_{liquid} V^2_{manifold} = P_{orifice} + \tfrac{1}{2}\rho_{liquid} V^2_{orifice} = constant \quad \text{App. 3. 9}$$

As the liquid leaves the injector orifices, it breaks into droplets without changing pressure, so $P_{orifices}$ equals combustion chamber pressure.

Substituting for the velocity of the liquid from a rearrangement of the mass continuity equation (chapter 1):

$$V = \frac{\dot{m}_{liquid}}{\rho_{liquid} A} \quad \text{App. 3. 10}$$

where $\dot{m}_{liquid}$ is the mass flow rate of liquid, A is the cross-sectional area of either the manifold or injector orifice, and ρ_{liquid} is the liquid density

gives:

$$P_{orifice} - P_{manifold} = \Delta P = \frac{\rho_{liquid} \dot{m}^2_{liquid}}{2\rho^2_{liquid}} \left(\frac{C}{(NA_{orifice})^2} - \frac{1}{A^2_{manifold}} \right) \quad \text{App. 3. 11}$$

$$= \frac{\dot{m}^2_{liquid}}{2\rho_{liquid}} \left(\frac{C}{(NA_{orifice})^2} - \frac{1}{A^2_{manifold}} \right) \quad \text{App. 3. 12}$$

where N is the number of orifices.

Note the introduction of a loss coefficient C. This represents the loss of total pressure (an energy loss) due to viscous losses/turbulence as the flow flows through the edges of the orifice. It depends on the fluid used and also how sharp the edges of the orifices are.

As the static pressure of the Nitrous liquid drops as it passes through the orifices, it vaporizes. This means that what flows through the injector is a foam of liquid and bubbles, but mostly vapour, and so traditional tables of loss coefficients or discharge coefficients don't work for this mixed fluid. You have to tailor this C coefficient until the time taken to empty the tank matches your test results.

We've found that a good starting value for C is 2.0 for Nitrous oxide. Interestingly, this value of 2.0 applies to either liquid or vapour Nitrous, because the liquid vaporizes almost completely into vapour as it flows through the orifices due to its pressure dropping significantly.

Rearranging, and assuming that $A^2_{manifold}$ is 'infinitely' larger than an injector orifice area gives:

$$\dot{m}_{liquid} = NA_{orifice} \sqrt{\frac{2\rho_{liquid} \Delta P}{C}} \quad \text{App. 3. 13}$$

Rearranging gives us the number of orifices required for a particular mass flow rate:

$$N = \frac{\dot{m}_{liquid}}{A_{orifice}\sqrt{\frac{2\rho_{liquid}\Delta P}{C}}} \quad \text{App. 3. 14}$$

Stagnation temperature

As I said in chapter 1, if you stick a temperature probe into the nozzle flow you'll get the stagnation temperature instead, which is also the temperature in the combustion chamber.

Using specific enthalpy and specific kinetic energy, we can relate the enthalpy h at some point of interest down the nozzle to the stagnation enthalpy h_0:

$$h_0 = h + \frac{V^2}{2} \quad \text{App. 3. 15}$$

where **enthalpy** = $C_p T$ App. 3. 16

then:

$$C_p T_0 = C_p T + \frac{V^2}{2} \quad \text{App. 3. 17}$$

dividing:

$$\frac{C_p T_0}{C_p T} = 1 + \frac{V^2}{2C_p T} \quad \text{App. 3. 18}$$

Now from the speed of sound equation App.3. 7 above:

$$V = Ma = M\sqrt{\gamma R T} \quad \text{App. 3. 19}$$

so:

$$\frac{T_0}{T} = 1 + \frac{M^2 \gamma R T}{2C_p T} \quad \text{App. 3. 20}$$

Now from chapter 1, γ is defined as:

$$\gamma = \frac{C_p}{C_v} \quad \text{App. 3. 21}$$

where γ is the isentropic exponent (ratio of specific heats) for the exhaust gas.

Also, R is defined as:

$$R = C_p - C_v \quad \text{App. 3. 22}$$

so:

$$\gamma = \frac{C_p}{C_v} = \frac{C_p}{C_p - R} \quad \text{App. 3. 23}$$

so:

$$\frac{1}{\gamma} = 1 - \frac{R}{C_p} \quad \text{App. 3.24}$$

then:

$$\frac{R}{C_p} = \frac{\gamma-1}{\gamma} \quad \text{App. 3.25}$$

Therefore substituting in App 3.19:

$$\frac{T_0}{T} = 1 + \frac{M^2\gamma}{2}\left(\frac{\gamma-1}{\gamma}\right) = 1 + \left(\frac{\gamma-1}{2}\right)M^2 \quad \text{App. 3.26}$$

Where M is the Mach number of the flow at the point down the nozzle where you're trying to measure the temperature.

This equation will also tell you how the temperature drops as the Mach number increases down the nozzle, because the stagnation temperature is just the temperature of the stagnant gas in the combustion chamber.

The isentropic flow relations

If we assume that the nozzle flow process can be approximated as being isentropic (constant entropy) then from chapter 1:

$$\Delta s = C_v ln\left(\frac{T_2}{T_1}\right) + R\, ln\left(\frac{v_2}{v_1}\right) = 0 \quad \text{App. 3.27}$$

where v is specific volume (the reciprocal of density)

so:

$$ln\left(\frac{T_2}{T_1}\right) = -\frac{R}{C_v} ln\left(\frac{v_2}{v_1}\right) = -\frac{R}{C_v} ln\left(\frac{\rho_1}{\rho_2}\right) \quad \text{App. 3.28}$$

as: **density** $\rho = \frac{1}{v}$

Now we saw in App. 3.21 that $\gamma = \frac{C_p}{C_v}$ and App. 3.22 that $R = C_p - C_v$

so:

$$\gamma = \frac{C_p}{C_v} = \frac{C_v + R}{C_v} = 1 + \frac{R}{C_v} \quad \text{App. 3.29}$$

so:

$$\frac{R}{C_v} = \gamma - 1 \quad \text{App. 3.30}$$

giving:

$$ln\left(\frac{T_2}{T_1}\right) = -(\gamma - 1) ln\left(\frac{\rho_1}{\rho_2}\right) \quad \text{App. 3.31}$$

Therefore from the rules of logarithms:

$$ln\left(\frac{T_2}{T_1}\right) = -ln\left(\frac{\rho_1}{\rho_2}\right)^{\gamma-1} = ln\left(\frac{\rho_2}{\rho_1}\right)^{\gamma-1} \quad \text{App. 3. 32}$$

so:

$$\left(\frac{T_2}{T_1}\right) = \left(\frac{\rho_2}{\rho_1}\right)^{\gamma-1} \quad \text{App. 3. 33}$$

Alternatively, from the Ideal gas law (see chapter 1 equation 1.14):

$$T = \frac{P}{\rho R} \quad \text{App. 3. 34}$$

so:

$$\left(\frac{T_2}{T_1}\right) = \frac{R\left(\frac{P_2}{\rho_2}\right)}{R\left(\frac{P_1}{\rho_1}\right)} = \frac{P_2}{P_1}\left(\frac{\rho_2}{\rho_1}\right)^{-1} \quad \text{App. 3. 35}$$

therefore:

$$\left(\frac{P_2}{P_1}\right)\left(\frac{\rho_2}{\rho_1}\right)^{-1} = \left(\frac{\rho_2}{\rho_1}\right)^{\gamma-1} \quad \text{App. 3. 36} \quad \text{(from App 3. 33)}$$

and on dividing:

$$\left(\frac{P_2}{P_1}\right) = \left(\frac{\rho_2}{\rho_1}\right)^{(\gamma-1)-(-1)} = \left(\frac{\rho_2}{\rho_1}\right)^{\gamma} \quad \text{App. 3. 37}$$

Finally, combining App 3. 33 and App 3. 37:

$$\left(\frac{T_2}{T_1}\right)^{\frac{1}{\gamma-1}} = \left(\frac{\rho_2}{\rho_1}\right) = \left(\frac{P_2}{P_1}\right)^{\frac{1}{\gamma}} \quad \text{App. 3. 38}$$

or:

$$\left(\frac{P_2}{P_1}\right) = \left(\frac{T_2}{T_1}\right)^{\frac{\gamma}{\gamma-1}} \quad \text{App. 3. 39}$$

If position 1 is taken as the combustion chamber, and postion 2 the nozzle exit, then these isentropic flow equations can be used to work out the nozzle exit temperature T_{exit} once you know the pressures at either end of the nozzle and the isentropic exponent (ratio of specific heats) γ for the exhaust gas:

$$T_{exit} = T_{chamber}\left(\frac{P_{exit}}{P_{chamber}}\right)^{\frac{\gamma-1}{\gamma}} \quad \text{App. 3. 40} \quad \text{(from App.3. 39)}$$

or:

$$T_{exit} = T_{chamber}\left(\frac{\rho_{exit}}{\rho_{chamber}}\right)^{\gamma-1} \quad \text{App. 3. 41} \quad \text{(from App. 3. 33)}$$

Alternatively, you can use the stagnation temperature equation App.3. 26 above to work out how the temperature drops as the Mach number down the nozzle increases. So you can then work out how the pressure and density change down the nozzle. As these pressure and density equations use the stagnation temperature equation, then they change only with Mach number as I said in chapter 4:

$$P = P_{chamber}\left(\frac{T}{T_{chamber}}\right)^{\frac{\gamma}{\gamma-1}} = P_{chamber}\left(\frac{1}{1+\left(\frac{\gamma-1}{2}\right)M^2}\right)^{\frac{\gamma}{\gamma-1}} \quad \text{App. 3. 42} \quad \text{(from App.3. 39)}$$

$$= P_{chamber}\left(1+\left(\frac{\gamma-1}{2}\right)M^2\right)^{\frac{-\gamma}{\gamma-1}} \quad \text{App. 3. 43}$$

similarly:

$$\rho = \rho_{chamber}\left(\frac{T}{T_{chamber}}\right)^{\frac{1}{\gamma-1}} = \rho_{chamber}\left(1+\left(\frac{\gamma-1}{2}\right)M^2\right)^{\frac{-1}{\gamma-1}} \quad \text{App. 3. 44} \quad \text{(from App.3. 33)}$$

When you come across a bunch of γ's (as you will in rocketry equations) it's best to work the γ's bit out first before doing the rest of the sum.

So in the isentropic equations above, if $\gamma = 1.257$ (from appendix 2)

then:

$$\frac{\gamma-1}{\gamma} = \frac{1.257-1}{1.257} = 0.2045 \quad \text{App. 3. 45}$$

Ideal (no pressure thrust) area ratio

The ideal nozzle area ratio ϵ (the nozzle size that brings the exhaust gasses down in pressure to exactly the value of the atmosphere's pressure outside) is derived thus:

From the mass continuity equation (see chapter 1), then for any two points in the nozzle:

$$\rho_1 A_1 V_1 = \rho_2 A_2 V_2 \quad \text{App. 3. 46}$$

Rearranging:

$$\frac{A_2}{A_1} = \frac{\rho_1 V_1}{\rho_2 V_2} \quad \text{App. 3. 47}$$

From the speed of sound equation App. 3. 7 above:

$$V = Ma = M\sqrt{\gamma R T} \quad \text{App. 3. 48}$$

so:

$$\frac{A_2}{A_1} = \frac{\rho_1 M_1 \sqrt{\gamma R T_1}}{\rho_2 M_2 \sqrt{\gamma R T_2}} = \frac{\rho_1 M_1 \sqrt{T_1}}{\rho_2 M_2 \sqrt{T_2}} = \left(\frac{\rho_1}{\rho_2}\right)\left(\frac{M_1}{M_2}\right)\sqrt{\frac{T_1}{T_2}} \quad \text{App. 3. 49}$$

Using the stagnation equation App 3. 26 and the isentropic flow relations App 3. 33 and App 3. 44:

$$\frac{A_2}{A_1} = \left(\frac{\rho_0\left(1+\left(\frac{\gamma-1}{2}\right)M_1^2\right)^{\frac{-1}{\gamma-1}}}{\rho_0\left(1+\left(\frac{\gamma-1}{2}\right)M_2^2\right)^{\frac{-1}{\gamma-1}}}\right) \left(\frac{M_1}{M_2}\right) \sqrt{\frac{T_0\left(1+\left(\frac{\gamma-1}{2}\right)M_1^2\right)^{-1}}{T_0\left(1+\left(\frac{\gamma-1}{2}\right)M_2^2\right)^{-1}}} = \left(\frac{M_1}{M_2}\right) \frac{\left(1+\left(\frac{\gamma-1}{2}\right)M_2^2\right)^{\frac{2+(\gamma-1)}{2(\gamma-1)}}}{\left(1+\left(\frac{\gamma-1}{2}\right)M_1^2\right)^{\frac{2+(\gamma-1)}{2(\gamma-1)}}} \quad \text{App. 3. 50}$$

$$= \left(\frac{M_1}{M_2}\right) \sqrt{\left(\frac{1+\left(\frac{\gamma-1}{2}\right)M_2^2}{1+\left(\frac{\gamma-1}{2}\right)M_1^2}\right)^{\frac{\gamma+1}{\gamma-1}}} \quad \text{App. 3. 51}$$

If we're interested in the area ratio, then A_2 = the nozzle exit A_e, and A_1 is the throat A^* where $M^*=1$ so:

$$1 + \left(\frac{\gamma-1}{2}\right)M_1^2 = 1 + \left(\frac{\gamma-1}{2}\right)1^2 = \frac{\gamma+1}{2} \quad \text{App. 3. 52}$$

therefore:

$$\epsilon = \frac{A_{exit}}{A^*} = \left(\frac{1}{M_{exit}}\right) \sqrt{\left[\left(\frac{2}{\gamma+1}\right)\left(1 + \left(\frac{\gamma-1}{2}\right)M_{exit}^2\right)\right]^{\frac{\gamma+1}{\gamma-1}}} \quad \text{App. 3. 53}$$

So the area ratio depends only on the nozzle exit Mach number.

Nozzle exit Mach number and exhaust velocity, M_{exit}, V_e

To get the Mach number and velocity at the nozzle exit, we use the specific enthalpy relation derived in chapter 5 equation 5.4:

$$V_e = \sqrt{2 C_p (T_{chamber} - T_{exit})} \quad \text{App. 3. 54}$$

Now we saw in App. 3. 21 that $\gamma = \frac{C_p}{C_v}$ and App. 3. 22 that $R = C_p - C_v$

so:

$$\frac{1}{\gamma} = \frac{C_v}{C_p} = \frac{C_p - R}{C_p} = 1 - \frac{R}{C_p} \quad \text{App. 3. 55}$$

so:

$$\frac{R}{C_p} = 1 - \frac{1}{\gamma} = \frac{\gamma-1}{\gamma} \quad \text{App. 3. 56}$$

rearranging:

$$C_p = R\left(\frac{\gamma}{\gamma-1}\right) \quad \text{App. 3. 57}$$

So substituting in App. 3. 54:

$$V_e = \sqrt{\left(\frac{2R\gamma}{\gamma-1}\right) T_{chamber}\left(1 - \frac{T_{exit}}{T_{chamber}}\right)} = \sqrt{\left(\frac{2R\gamma}{\gamma-1}\right) T_{chamber}\left[1 - \left(\frac{P_{exit}}{P_{chamber}}\right)^{\frac{\gamma-1}{\gamma}}\right]} \quad \text{App. 3. 58}$$

using the isentropic flow equation App.3. 39.

By a similar analysis, we can express V_e as:

$$V_e = \sqrt{\left(\frac{2R\gamma}{\gamma-1}\right) T_{exit}\left(\frac{T_{chamber}}{T_{exit}} - 1\right)} = \sqrt{\left(\frac{2R\gamma}{\gamma-1}\right) T_{exit}\left[\left(\frac{P_{chamber}}{P_{exit}}\right)^{\frac{\gamma-1}{\gamma}} - 1\right]} \quad \text{App. 3. 59}$$

Now from App. 3. 7 the speed of sound $a = \sqrt{\gamma R T}$

so:

$$V_e = \sqrt{\left(\frac{2R\gamma}{\gamma-1}\right) T_{exit}\left[\left(\frac{P_{chamber}}{P_{exit}}\right)^{\frac{\gamma-1}{\gamma}} - 1\right]} = a_{exit}\sqrt{\left(\frac{2}{\gamma-1}\right)\left[\left(\frac{P_{chamber}}{P_{exit}}\right)^{\frac{\gamma-1}{\gamma}} - 1\right]} \quad \text{App. 3. 60}$$

dividing both sides by a_{exit} and using equation App. 3. 8:

$$M_{exit} = \sqrt{\left(\frac{2}{\gamma-1}\right)\left[\left(\frac{P_{chamber}}{P_{exit}}\right)^{\frac{\gamma-1}{\gamma}} - 1\right]} \quad \text{App. 3. 61}$$

Characteristic velocity C*

Starting with the mass continuity equation at the throat: $\dot{m}_{nozzle} = \rho^* A^* V^*$

and using the Ideal gas equation: $\rho^* = \frac{P^*}{R T^*}$

and we know that at the throat, the gas velocity is at Mach 1, the speed of sound,

so:

$$V^* = \sqrt{\gamma R T^*} \quad \text{App. 3. 62}$$

from the speed of sound equation App.3. 7 above.

therefore:

$$\dot{m}_{nozzle} = \frac{P^*}{R T^*} A^* \sqrt{\gamma R T^*} = P^* A^* \sqrt{\frac{\gamma}{R T^*}} \quad \text{App. 3. 63}$$

So $\dot{m}_{nozzle}$ depends inversely on the square root of the gas temperature at the throat as I pointed out earlier.

We would prefer to use combustion chamber values of pressure and temperature rather than throat values.

Now in the combustion chamber, the flow is essentially stagnant, so stagnation conditions prevail. So using the stagnation temperature equation (remembering that Mach number M is 1 at the throat) and the isentropic flow relations:

$$\frac{T_{chamber}}{T^*} = 1 + \left(\frac{\gamma-1}{2}\right)1^2 = \frac{\gamma+1}{2} \qquad \text{App. 3. 64} \quad \text{(from App. 3. 26)}$$

and so:

$$\frac{P_{chamber}}{P^*} = \left(\frac{T_{chamber}}{T^*}\right)^{\frac{\gamma}{\gamma-1}} = \left(\frac{\gamma+1}{2}\right)^{\frac{\gamma}{\gamma-1}} \qquad \text{App. 3. 65} \quad \text{(from App. 3. 39)}$$

So substituting in App. 3. 63:

$$\dot{m}_{nozzle} = P_{chamber}\left(\frac{\gamma+1}{2}\right)^{\frac{-\gamma}{\gamma-1}} A^* \sqrt{\frac{\gamma}{R\,T_{chamber}\left(\frac{\gamma+1}{2}\right)^{-1}}} \qquad \text{App. 3. 66}$$

$$= \frac{P_{chamber}\,A^*}{\sqrt{R\,T_{chamber}}}\sqrt{\gamma}\left(\frac{\gamma+1}{2}\right)^{\left(\frac{1}{2}-\frac{\gamma}{\gamma-1}\right)} = \frac{P_{chamber}\,A^*}{\sqrt{R\,T_{chamber}}}\sqrt{\gamma}\left(\frac{\gamma+1}{2}\right)^{\frac{(\gamma-1)-2\gamma}{2(\gamma-1)}} \qquad \text{App. 3. 67}$$

$$= \frac{P_{chamber}\,A^*}{\sqrt{R\,T_{chamber}}}\sqrt{\gamma}\left(\frac{2}{\gamma+1}\right)^{\left[\frac{1}{2}\left(\frac{\gamma+1}{\gamma-1}\right)\right]} = \frac{P_{chamber}\,A^*}{\sqrt{R\,T_{chamber}}}\sqrt{\Gamma} \qquad \text{App. 3. 68}$$

$$= \frac{P_{chamber}\,A^*}{\sqrt{\frac{R\,T_{chamber}}{\Gamma}}} = \frac{P_{chamber}\,A^*}{C^*} \qquad \text{by defining } C^* \text{ as: } \sqrt{\frac{R\,T_{chamber}}{\Gamma}} \qquad \text{App. 3. 69}$$

where:

$$\Gamma = \gamma\left(\frac{2}{\gamma+1}\right)^{\left(\frac{\gamma+1}{\gamma-1}\right)} \qquad \text{App. 3. 70}$$

Thrust coefficient C_F

Using the above derivation for V_e (App. 3. 58)

$$V_e = \sqrt{\left(\frac{2\,R\,\gamma}{\gamma-1}\right)T_{chamber}\left[1 - \left(\frac{P_{exit}}{P_{chamber}}\right)^{\frac{\gamma-1}{\gamma}}\right]}$$

and from chapter 5: $\textbf{Thrust} = \dot{m}V_e + (P_{exit} - P_a)A_{exit}$

we get:

$$\textbf{thrust} = \dot{m}\sqrt{\left(\frac{2\,R\,\gamma}{\gamma-1}\right)T_{chamber}\left[1 - \left(\frac{P_{exit}}{P_{chamber}}\right)^{\frac{\gamma-1}{\gamma}}\right]} + (P_{exit} - P_a)A_{exit} \qquad \text{App. 3. 71}$$

Now, we'd like to express $\dot{m} = \rho A V$ also in terms of chamber conditions, but there's a problem because V is assumed zero (stagnation assumption) in the chamber, which would drive A to infinity to get a finite mass flow rate.

So we'll use throat conditions initially:

$$\dot{m} = \rho^* A^* V^* \quad \text{App. 3. 72}$$

Using the stagnation equation App 3. 26 and knowing that the flow speed is Mach 1 at the throat:

$$\frac{T_{chamber}}{T^*} = 1 + \left(\frac{\gamma-1}{2}\right) 1^2 = \frac{\gamma+1}{2} \quad \text{App. 3. 73}$$

and so from the isentropic relation App. 3. 33:

$$\frac{\rho_{chamber}}{\rho^*} = \left(\frac{T_{chamber}}{T^*}\right)^{\frac{1}{\gamma-1}} = \left(\frac{\gamma+1}{2}\right)^{\frac{1}{\gamma-1}} \quad \text{App. 3. 74}$$

Also, we know that at the throat:

$$V^* = \sqrt{\gamma R T^*} \quad \text{App. 3. 75}$$

from the speed of sound equation App. 3. 7 above.

So using very similar algebra to the above derivation of C^*:

$$thrust = \rho_{chamber} A^* \sqrt{\gamma R T_{chamber}} \sqrt{\left(\frac{2}{\gamma+1}\right)^{\frac{\gamma+1}{\gamma-1}}} \sqrt{\left(\frac{2R\gamma}{\gamma-1}\right) T_{chamber} \left[1 - \left(\frac{P_{exit}}{P_{chamber}}\right)^{\frac{\gamma-1}{\gamma}}\right]}$$

$$+ (P_{exit} - P_a) \quad \text{App. 3. 76}$$

$$= (\rho_{chamber} R T_{chamber}) A^* \sqrt{\gamma} \sqrt{\left(\frac{2}{\gamma+1}\right)^{\frac{\gamma+1}{\gamma-1}}} \sqrt{\left(\frac{2\gamma}{\gamma-1}\right) \left[1 - \left(\frac{P_{exit}}{P_{chamber}}\right)^{\frac{\gamma-1}{\gamma}}\right]}$$

$$+ (P_{exit} - P_a) A_{exit} \quad \text{App. 3. 77}$$

Using the Ideal gas law $P = \rho R T$: (equation 1.14)

$$thrust = P_{chamber} A^* \sqrt{\Gamma \left(\frac{2\gamma}{\gamma-1}\right) \left[1 - \left(\frac{P_{exit}}{P_{chamber}}\right)^{\frac{\gamma-1}{\gamma}}\right]} + (P_{exit} - P_a) A_{exit} \quad \text{App. 3. 78}$$

Now, from the definition of C_F:

$$C_F = \frac{thrust}{A^* P_{chamber}} \quad \text{App. 3. 79}$$

then:

$$C_F = \sqrt{\Gamma\left(\frac{2\gamma}{\gamma-1}\right)\left(1-\left(\frac{P_{exit}}{P_{chamber}}\right)^{\frac{\gamma-1}{\gamma}}\right)} + \left(\frac{A_{exit}}{P_{chamber}\,A^*}\right)(P_{exit}-P_a) \qquad \text{App. 3. 80}$$

In both the C^* and C_F equations:

$$\Gamma = \gamma\left(\frac{2}{\gamma+1}\right)^{\left(\frac{\gamma+1}{\gamma-1}\right)} \qquad \text{App. 3. 81}$$

and is a fixed number.

Note that some books use a different equation for Γ but they're putting it into slightly different equations for C^* and C_F. The final answers work out the same though.

If $\gamma = 1.257$ (from appendix 2) then $\Gamma = 0.43477$

Appendix 4: Altering the PROPEP 3 input file

The PROPEP 3 software is extremely useful. It's fast and it's fairly easy to add new propellants to the end of its very long list of possible rocket propellants.

PROPEP 3 is available to download at:

http://www.tclogger.com courtesy of Dave Cooper (click on the 'PROPEP 3 now available' icon).

PROPEP 3 can simulate burning with Nitrous gas (vapour) but can't sim burning with Nitrous liquid as it doesn't have any data on liquid Nitrous. You have to alter the software to get it to do that.

PROPEP 3, like the original PROPEP software, reads an input file called 'pepcoded.daf' which is a long list of propellants. If you want to sim liquid Nitrous, you'll have to alter this file.

I've made this alteration for you and put the altered pepcoded.daf on my rocketry society's website www.aspirespace.org.uk/technical_papers.html in a downloadable zip file called 'what the results file means.zip'

Once you've installed PROPEP 3, pepcoded.daf will appear in subdirectory '/documents/ProPEP 3' on your computer. Replace this file with the altered one.

I've also put a document explaining what the PROPEP 3 results file means on the Aspirespace website in the same zip file.

Before you do anything, read the text file 'read this first' (which is in the zip file) as it tells you how to set the software up and how to run it.

However, if you download pepcoded.daf from elsewhere, you'll need to make the Nitrous alteration yourself. The following tells you how to do this, perhaps you want to add your own ingredients.

Please make a copy of the pepcoded.daf file for safety before you start altering it just in case you make a mistake and cause PROPEP to crash.

Open pepcoded.daf using Microsoft Notepad, which you'll find in the 'accessories' folder on the Windows start menu. If you use a different program such as Wordpad or Word to edit this file, they leave extra invisible information in the file when you save it, which will make PROPEP crash.

So, using Notepad, scroll down the file until you see the line that has NITROUS OXIDE written on it. You'll see various numbers on this line; the second-last number should read 443 (if it doesn't, then make it 443.) This is the enthalpy of formation (also called the heat of formation, see chapter 6) of Nitrous oxide gas, which is the energy you need to put in to make Nitrous oxide from its elements of nitrogen and oxygen.

The calculation goes like this:

The molecular mass (relative molecular mass or molecular weight) of Nitrous oxide is 44.013 grams per mole (chapter 1).

The enthalpy of formation of Nitrous oxide gas is given in many textbooks as 81.6 kiloJoules per mole. (This is the number if Nitrous oxide is formed at 25 degrees C, which is the internationally agreed temperature to measure enthalpy of formation at.)

Now divide this enthalpy of formation by the molecular weight, and then multiply the answer by 1,000. This gives 1,854 Joules per gram. Finally, multiply by 0.2388459 to convert into calories per gram, which is the horrible unit that PROPEP uses. You'll get 443 calories per gram.

So that's the number for Nitrous gas; how do we get the number for Nitrous liquid?

It takes energy to convert a liquid into a gas.

This energy comes from the Nitrous itself, which then gets considerably colder. Inside the combustion chamber, some of the energy from the combustion is wasted warming the Nitrous up gain to room temperature (and then on to much higher temperatures) so that's why the exhaust velocity ends up being lower.

The energy required to vaporize the Nitrous is called the Heat (enthalpy) of vaporization (see chapter 1) and for Nitrous oxide (at the same 25 degrees C) it has a value of 146.5 kiloJoules per kilogram. Multiply this by 0.2388459 to convert to calories per gram to get 35 calories per gram.

Now we subtract this Heat of vaporization from the enthalpy of formation of Nitrous gas:

443 – 35 = 408 calories per gram. This is the value to put into PROPEP for Nitrous liquid.

Copy the line for Nitrous from pepcoded.daf, and paste it to the end of that file, calling it NITROUS (LIQUID). Replace the 443 by 408.

Make sure that the first number on the line is one number higher than the line before it; you'll see that each line has a number.

Make sure that your new line is exactly the same length as the other lines.

Now save the file.

Double-click on the PROPEP application to run the program. This puts the results in a text file called 'print'; the last two lines of this file are the ones you want.

For an explanation of the results read the pdf file 'What the PROPEP results mean' included in my zip file.

Note: some versions of pepcoded.daf that you may find on the internet incorrectly list polyethylene as:

866 POLYETHYLENE   2C 3H 0 0 0 0 -453 .0325]
867 POLYETHYLENE (FILM)   2C 3H 0 0 0 0 -491 .0325]
868 POLYETHYLENE (PELLETS) 2C 3H 0 0 0 0 -478 .0325]

when it should be *four* hydrogen atoms:

866 POLYETHYLENE   2C 4H 0 0 0 0 -453 .0325]
867 POLYETHYLENE (FILM)   2C 4H 0 0 0 0 -491 .0325]
868 POLYETHYLENE (PELLETS) 2C 4H 0 0 0 0 -478 .0325]

Appendix 5: Converting regression data

Experimental regression rate data can be given in many units. To convert regression rate data from one set of units to another, remember to take account of the power that the mass flux G is raised to.

For example:

$0.104 \times G^{0.352}$ **millimetres/second regression rate per gram/centimetre²second** mass flux

equals (as there are 10,000 square centimetres in a square metre):

$(0.104 / 1{,}000) \times (1{,}000 / 10{,}000)^{0.352} \times G^{0.352}$

$= 4.6242 \times 10^{-5} \times G^{0.352}$ **metres/second regression rate per kilogram/metre²second** mass flux

Appendix 6: Plotting a bell nozzle contour

Rao in America and Shmyglevsky in Russia found a way to design an optimum nozzle that was relatively short: it resembled a church bell and was hence known as a 'bell nozzle'.

For ease of use, Rao measured the lengths of his resulting nozzles as fractions of the length of a standard 15 degree half-angle conical nozzle which has length:

$$L_{N\_cone} = \frac{(\sqrt{\epsilon}-1) R_t}{\tan(15)} \quad \textbf{App.6. 1} \quad \text{where } R_t \text{ is the radius of the throat.}$$

So an '80% bell' would have a length of 0.8 times this length.

Rao was clearly also schooled in traditional geometry. He was able to spot that the bell part of his bell nozzles could be approximated very closely by a skewed parabola allowing us to quickly sketch his nozzles with negligible loss of thrust performance.

These sketched approximations are known as **Thrust Optimised Parabolic nozzles** (TOP), and have found use on a variety of actual launch vehicles because they perform better when over-expanded (see page 65) at ground-level altitude than the actual optimised bell nozzle (flow separation from the TOP nozzle wall is delayed at high back-pressure).

Rao's parabolas are known in Europe as **quadratic Bézier curves**, after French engineer Pierre Bézier. Bézier curves are now used extensively in computer graphics.

The shape of the bell nozzle changes only minutely with the propellants used (varying ratio of specific heats γ) so one TOP nozzle methodology fits all propellants and is described below.

Construction

The 'Rao nozzle' starts with Rao's preferred throat geometry:

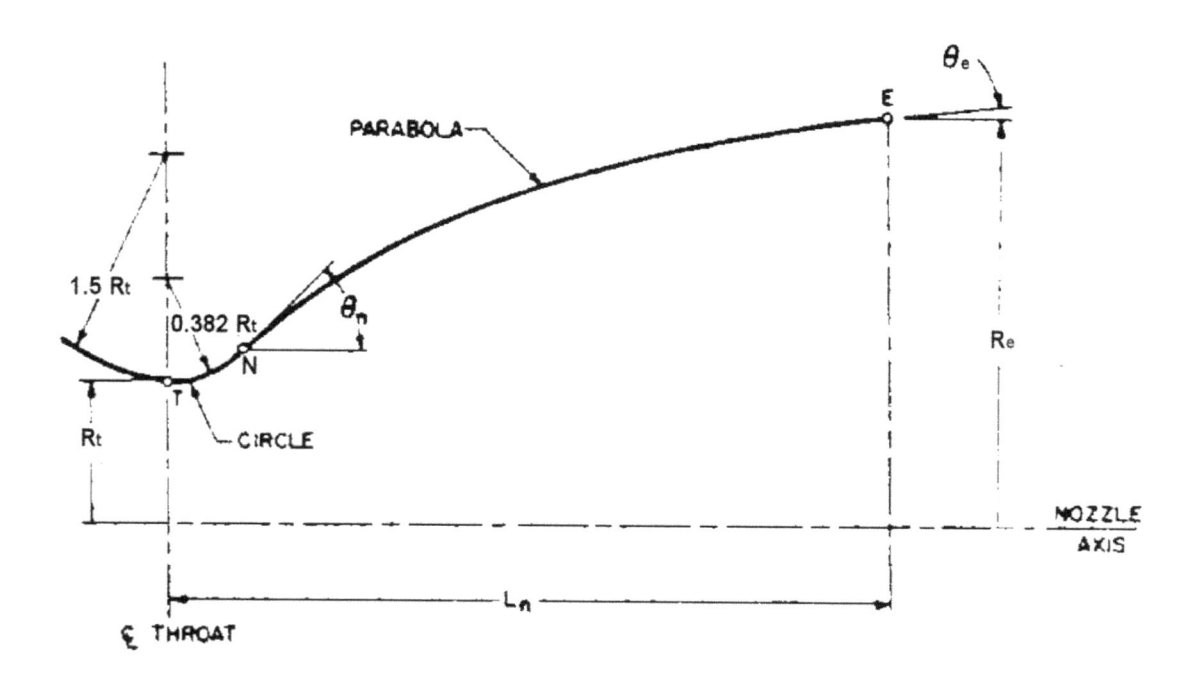

Two circular arcs are drawn (see previous sketch):

The first curve, of radius 1.5 R_t, is drawn from an angle of say, -135 degrees, to the throat at -90 degrees (angles measured from the arc's origin). Then the second curve of radius 0.382 R_t is drawn from this angle of -90 degrees to an angle of ($θn$ - 90) at inflection point N. ($θn$ is given on the next page, and R_t is the throat radius, R_e is the exit radius.)

Then a skewed parabola is drawn from point N to nozzle exit point E, tangent to the throat curve, and starting at an angle of $θn$ and ending at an angle of $θe$.

The radius of the nozzle exit: $R_e = \sqrt{\epsilon}\, R_t$ **App.6. 2**

and:

$$\text{nozzle length } L_N = 0.8 \left(\frac{(\sqrt{\epsilon}-1)\, R_t}{\tan(15)} \right) \quad \textbf{App.6.3} \quad \text{for an 80\% bell from equation App.6.1}$$

Angles $θn$ and $θe$ were pre-calculated by Rao to match his bell nozzle, and presented as graphical data from which the following chart is reproduced for various percent lengths:

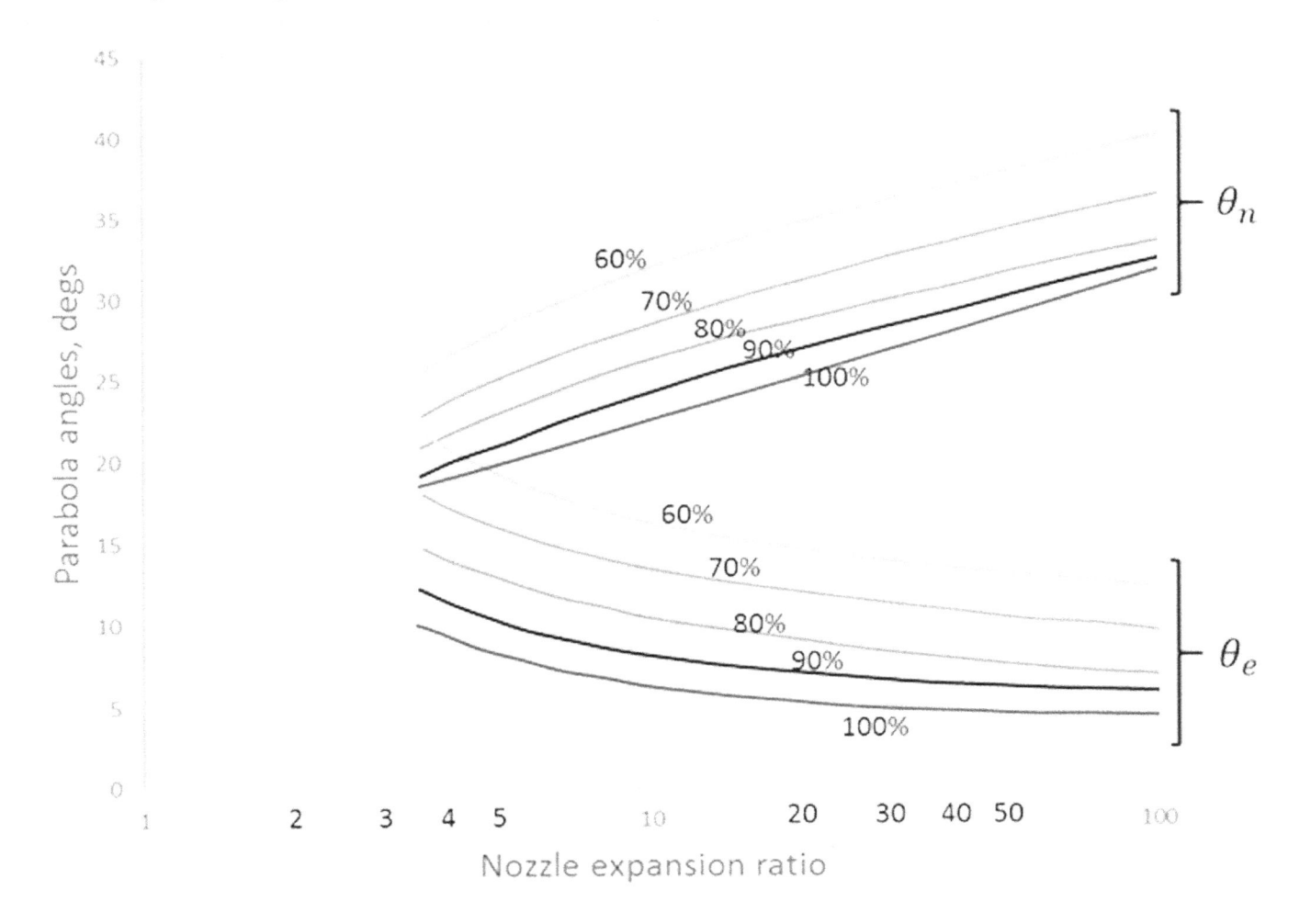

(The data for expansion ratios greater than 50 is extrapolated.)

A parabola is then sketched out using a very ancient geometrical method for drawing a parabola:

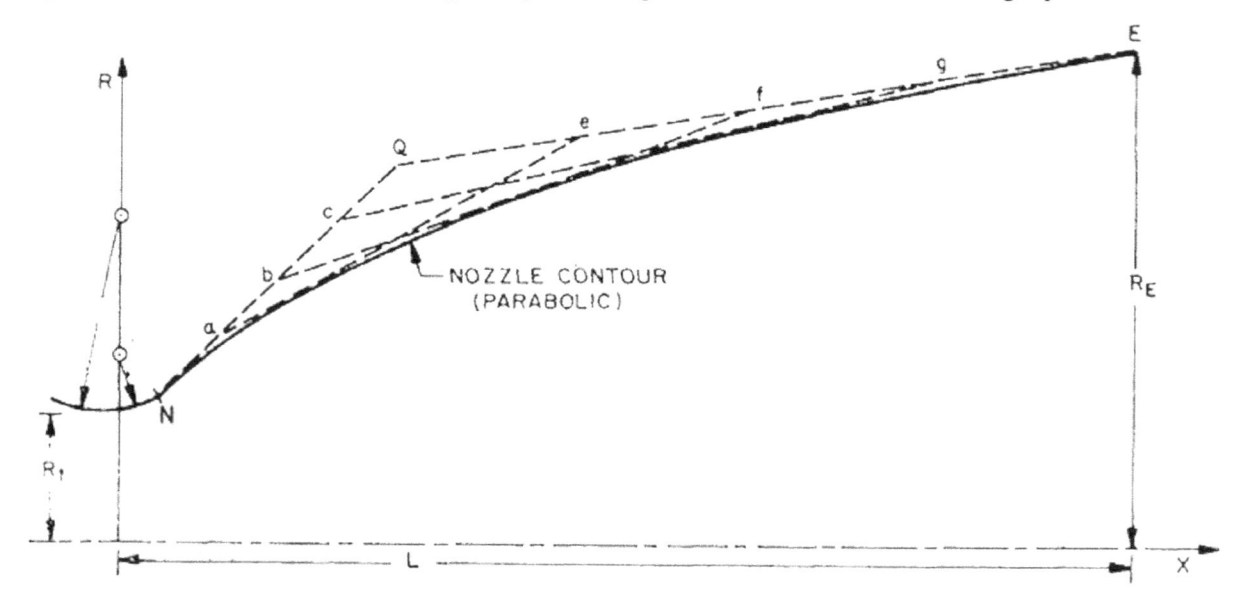

Straight lines are drawn at angles θn from point N, and θe back from point E, terminating where these lines cross at point Q.

Next, both of these lines are divided into an equal number of divisions.

These are labelled a,b,c and e,f,g.

A straight line is then drawn from point a to point e, then from b to f, and c to g. These form a mesh, the edge of which gives the parabola outline. The parabola is also tangent to the lines QN and QE.

Using many more divisions - on a CAD package for example - gives a sharper contour. Removing all but the bottom of the mesh gives a series of straight-line segments: joining the midpoint of each segment with a smooth curve such as a CAD spline gives the nozzle contour.

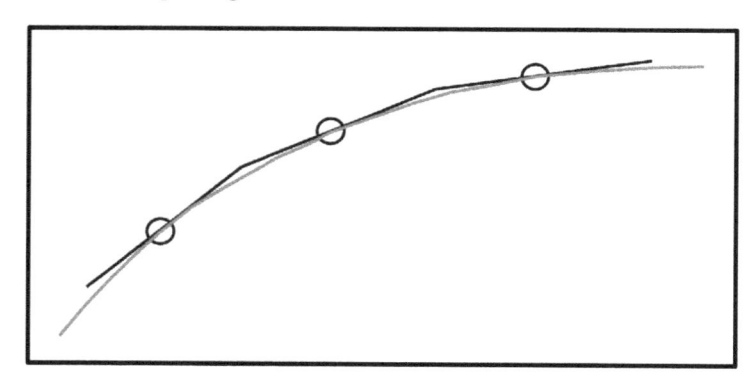

Alternatively, interpolating along the construction lines gives the same points. In our example with four divisions and (4-1) construction lines, the parabola is defined at 1/4 the distance along line a-e, 2/4 along line b-f, and 3/4 along c-f.

Efficiency

At a length ratio of 85% bell, a thrust coefficient nozzle efficiency of 99% is reached, and only 0.2% of additional performance can be gained by increasing the length ratio to 100%. For this reason, 85% is often taken as the upper bound. At length ratios below 70%, nozzle efficiency suffers. For these reasons, the 80% bell parabola is often chosen.

Example

An 80% bell nozzle with an area ratio of 70 - the expansion ratio we chose for our Space nozzle - has angles $\theta n = 33°$ and $\theta e = 7°$ from the above chart. Drawn on a CAD package it looks like this (mesh removed for clarity):

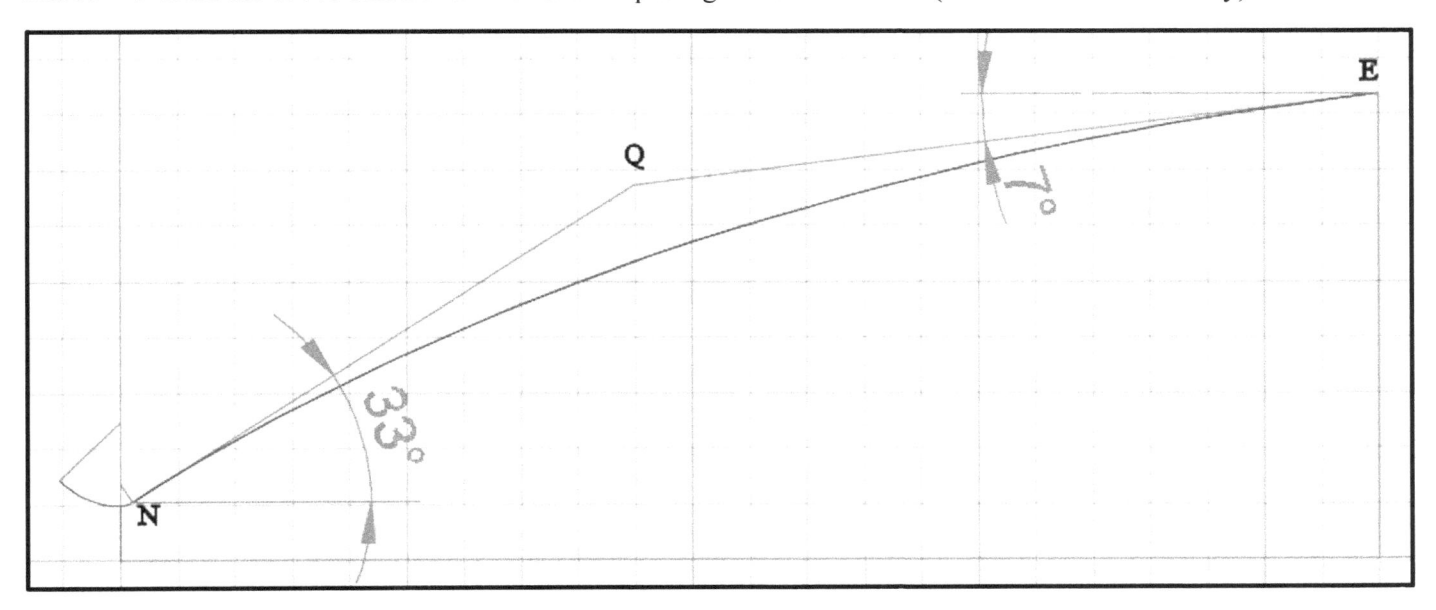

Alternatively, a mathematical approach can be taken to plot the curves.

The throat

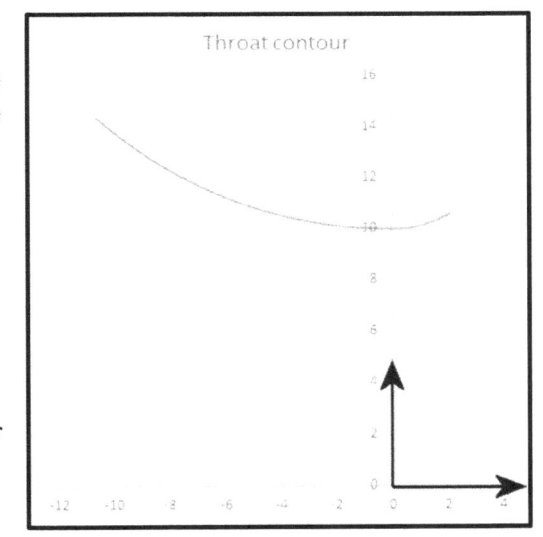

The equations of the above circular arcs constructing the throat are defined trigonometrically, defining the origin of the coordinates as the centre of the narrowest part of the throat:

For the entrant section:

$$x = 1.5\, R_t \cos\theta$$

$$y = 1.5\, R_t \sin\theta + 1.5\, R_t + R_t \quad \text{App.6. 4}$$

where: $-135 \leq \theta \leq -90$

(The initial angle isn't defined and is up to the combustion chamber designer, -135 degrees is typical.)

For the exit section:

$$x = 0.382\, R_t \cos\theta$$

$$y = 0.382\, R_t \sin\theta + 0.382\, R_t + R_t \quad \text{App.6. 5}$$

where: $-90 \leq \theta \leq (\theta_n - 90)$

The bell

The bell is a quadratic Bézier curve, which has equations:

$$x(t) = (1-t)^2 N_x + 2(1-t)t\, Q_x + t^2 E_x \quad 0 \leq t \leq 1$$

$$y(t) = (1-t)^2 N_y + 2(1-t)t\, Q_y + t^2 E_y \quad 0 \leq t \leq 1 \quad \text{App.6. 6}$$

Selecting equally spaced divisions between 0 and 1 produces the points described earlier in the graphical method, for example 0.25, 0.5, and 0.75.

Equations App.6.6 are defined by points N, Q, and E (see the graphical method earlier for the locations of these points).

Point N is defined by equations App.6.5 setting the angle to $(\theta n - 90)$.

Coordinate E_x is defined by equation App.6.3, and coordinate E_y is defined by equation App.6.2.

Point Q is the intersection of the lines:

$$\overrightarrow{NQ} = m_1 x + C_1 \text{ and: } \overrightarrow{QE} = m_2 x + C_2 \quad \text{App.6. 7}$$

where:

$$\text{gradient } m_1 = \tan(\theta_n)\,,\ \text{gradient } m_2 = \tan(\theta_e) \quad \text{App.6. 8}$$

and:

$$\text{intercept } C_1 = N_y - m_1 N_x\,,\ \text{intercept } C_2 = E_y - m_2 E_x \quad \text{App.6. 9}$$

The intersection of these two lines (at point Q) is given by:

$$Q_x = \frac{(C_2 - C_1)}{(m_1 - m_2)},\ Q_y = \frac{(m_1 C_2 - m_2 C_1)}{(m_1 - m_2)} \quad \text{App.6. 10}$$

Caution

A word of warning: bell nozzles are optimised for combustion chambers that maintain a fixed pressure. Nitrous hybrid chambers lose pressure throughout the burn, so it's possible that some other kind of nozzle geometry would be more efficient over the whole burn.

References

1. "University Physics 6th edition", (metric edition), Sears, Zemansky, and Young, Addison Wesley world student series ISBN 0-201-07199-1

2. "Space Propulsion Analysis and Design", Ronald .W. Humble, Gary .N. Henry and Wiley J. Larson, McGraw Hill Space Technology Series ISBN 0-07-031320-2

3. "Engineering Thermodynamics, Work, and Heat transfer (S.I. units) 4th edition", Rogers and Mayhew, Prentice Hall ISBN 0-582-04566-5

4. http://web.mit.edu/16.unified/www/FALL/thermodynamics/thermo_6.htm

5. "Thermophysical properties of Nitrous oxide", Engineering Sciences Data Unit (ESDU) sheet 91022

6. "Volumetric Behavior of Nitrous Oxide. Pressure-Volume Isotherms at High Pressures", E. J. Couch, K. A. Kobe, Journal of Chemical Engineering Data, 6 (2), pp 229–233, 1961

7. Communication with Dr Bruce P. Dunn, University of British Columbia and Dunn Engineering, Several articles on self-pressurised peroxide rockets and experiments on propane tanks, as well as email communications with the author on the subject of numerical modelling of the tank liquid emptying process.

8. "Rocket propulsion elements", Sutton, Biblarz, seventh edition, John Wiley and Sons, ISBN: 0-471-32642-9

9. "Design Of Liquid Propellant Rocket Engines", Huzel, D.K, and Huang, D.H, NASA SP-125, Scientific and Technical Information Division, Office of Technology Utilisation, National Aeronautics and Space Administration, Washington DC, USA (1967) (downloadable)

10. "Understanding Aerodynamics – arguing from the real physics", Doug McLean, John Wiley and sons Ltd. 2013, ISBN 978-1-119-96751-4

11. "Linus Pauling: the nature of the chemical bond"
    http://scarc.library.oregonstate.edu/coll/pauling/bond/narrative/page1.html

12. http://www.avogadro.co.uk/h_and_s/bondenthalpy/bondenthalpy.htm

13. http://sciencequestionswithsurprisinganswers.org/2013/06/27/when-does-the-breaking-of-chemical-bonds-release-energy/

14. "Chemwiki:"
    http://chemwiki.ucdavis.edu/Theoretical_Chemistry/Chemical_Bonding/General_Principles_of_Chemical_Bonding/Bond_Energies

15. "Ending Misconceptions About the Energy of Chemical Bonds",
    http://apcentral.collegeboard.com/apc/members/courses/teachers_corner/49039.html

16. "What is a chemical bond?", http://www.chem1.com/acad/webtext/chembond/cb01.html

17. "Chemistry for dummies", John T. Moore, 2nd edition, Wiley publishing Inc., 2011, ISBN: 978-1-118-00730-3

18. "Chemical Equilibrium",
    http://faculty.chem.queensu.ca/people/faculty/mombourquette/FirstYrChem/equilibrium/

19. "Gibbs free energy", https://www.chem.tamu.edu/class/majors/tutorialnotefiles/gibbs.htm

20. Aerocon hybrid factors paper 1997, and design manual 1995

21. Space Propulsion Group website, http://www.spg-corp.com

22. "Hydraulic diameter", Wikipedia

23. "Safety design for space systems", Musgrave, Larsen, Sgobba, International Association for the advancement of space safety, Elsevier publishing, ISBN 978-0-7506-8580-1

24. Advice from Rocket Services, Bere Regis, Dorset.

25. "Nitrous oxide trailer rupture", Report at CGA seminar 'Safety and reliability of industrial gasses, equipment and facilities' Oct 15-17, 2001, St Louis, Missouri

26. "Untersuchungen von Zerfallsfahigheit von Disticksoffoxid (Investigations on decomposability of Nitrous oxide)", D. Conrad, S.Deitlen Bundesanstalt fur Materialprufung (BAM) research report 89, Berlin 1983

27. "Dangers of Nitrous oxide no surprise", M. Holthaus, Space News article 3rd Sept. 2007

28. "Investigation of decomposition characteristics of gaseous and liquid Nitrous oxide", G. W. Rhodes, Air Force Weapons Laboratory, Kirtland airforce base, New Mexico, July 1974 (AD-784 802) (A compilation of results by Pratt and Whitney, and Rocketdyne)

29. "Modelling of N_2O decomposition events", Arif Karabeyoglu, Jonny Dyer, Jose Stevens, Brian Cantwell, Space Propulsion Group Inc (from their website)

30. "Handling considerations of Nitrous oxide in hybrid rocket motor testing", Zachary Thicksten, Frank Macklin, John Campbell, AIAA 2008-4830 44th Joint propulsion conference

31. Nitrous safety downloadable presentation, Bruno Berger, Swiss Propulsion Laboratory website http://www.spl.ch/

32. N20 safety guidelines downloadable pdf, Scaled Composites website

33. Air Liquide webpage: Nitrous compatibility chart (which is out of date).

34. CE 405: Design of Steel Structures – Prof. Dr. A. Varma
https://www.egr.msu.edu/~harichan/classes/ce405/chap5.pdf

35. "Roarks' Formulas for stress and strain, 7th edition", Young, Budynas, McGraw Hill international

36. Wikipedia: Von Mises yield criterion

37. United Kingdom Rocketry Association safety code, www.ukra.org.uk

38. Testing advice from several professional rocketeers, such as the late Ian Smith, Airborne Engineering, and Rocket Services, Dorset.

39. "Exhaust Nozzle Contour for Optimum Thrust", G.V.R. Rao, Jet Propulsion 28, 377 (1958)

40. "Liquid Rocket Engine Nozzles", NASA SP-8120, July 1976

41. http://www.wikihow.com/Draw-a-Parabolic-Curve-(a-Curve-with-Straight-Lines)

42. Wikipedia: quadratic Bezier curve

43. www.waters.to/blog/rocket-nozzles-part-1-the-math/

44. Wikipedia: line-line intersection

45. www.wikihow.com/Find-the-Equation-of-a-Line

46. Southampton University Aero/Astro course notes 'Astronautics' and 'Thermodynamics'

47. Open University physics course, 2nd year notes

Index

ablative, 146

absolute zero, 16

action, 12

action area, 12

adiabatic, 21

aerodynamic drag, 30

angle of attack, 161

Apogee, 30

area ratio, 66

atomised, 133

Avogadro's number, 16

back-pressure, 72

bell nozzle, 70

Bernoulii's equation, 167

bipropellant, 27

BLEVE, 111

blockhouse, 149

bond length, 84

boundary layer, 73, 96

boundary layer growth, 96

burnout, 25

burst disc, 103

carbon-carbon, 147

catalyst, 109

centre of gravity, 34, 159

centre of pressure, 160

Characteristic velocity (C*), 74

chemical bond, 82

choking, 61

Combustion, 12, 84

combustion chamber, 12

components, 15

compressibility factor, 17

condensation, 44

conical nozzle, 66

conserved, 19

covalent bond, 84

Critical point, 36

cryogenic, 35

data logging, 151

deflagration limit, 103

degrees Kelvin, 16

detonation, 26

differential calculus, 9

dip-tube, 43

doublet injector, 134

Drag coefficient, 31

drag loss, 31

dynamic pressure, 31

earth, 151

effective exhaust velocity, 71
elastic behaviour, 141
electromagnetic field, 85
electromagnetic radiation, 85
electrostatic force, 82
endothermic, 84
enthalpy, 21
enthalpy of formation, 87
enthalpy of the products., 87
enthalpy of the reactants, 87
equilibrium flow, 91
evaporation, 44
exhaust velocity, 54
exothermic, 84
expansion, 58
expansion ratio, 66
expulsion powder, 103
factored design strength, 144
fill tank, 42
first law of thermodynamics, 21
flame zone, 97
fluxion notation, 10
force, 11
Force coefficient (C_F), 74
forced convection, 97
frozen flow, 91

fuel, 25
Glass Transition Temperature, 97
graphite, 147
gravity loss, 31
gravity turn trajectory, 160
HDPE, 33
head, 42
heat of formation. *See* enthalpy of formation
heat of reaction, 87
heater charge, 102
hoop stress, 142
HPR (High power rocketry), 30
HTPB, 32
hydraulic diameter, 98
hydrocarbons, 85
hydro-test, 107
igniter, 25
inertia, 13
injector, 27
integral calculus, 10
internal energy, 20
ionic bond, 84
isentropic exponent, 20
isentropic flow, 24
isentropic flow relations, 24
jetavator, 162

Kármán line, 30

kilomole, 16

kinetic theory, 14

Laminar flow, 96

Large rockets, 30

latent heat of fusion. *See* specific heat of fusion

latent heat of vaporization. *See* specific heat of vaporisation

Law of conservation of momentum, 13

line of action, 11

liquid oxygen, 35

load cell, 67, 150

longitudinal stress, 142

mass, 11

mass continuity equation, 17

mass flow rate, 17, 54

Mass flux, 17

Mass ratio, 119, 120

mass-specific, 12

Maximum Expected Operating Pressure, 106

mixture ratio, 25

modulus of elasticity, 141

molecular mass, 16

molecule, 14

momentum, 13

momentum thrust., 56

multiport hybrid, 101

normal, 15

nozzle divergence angle, 93

nozzle throat, 12

nucleus, 83

numerical integration, 120

orifices, 27

O-rings, 148

over-expanded, 73

over-pressure, 26

oxidation, 85

oxidiser, 25

oxidises, 85

Pascals, 17

perfect gas law. *See* ideal gas law

phase change, 20

phase diagram, 39

phenolic resin, 146

photons, 86

Poisson's ratio, 146

port, 28

post-combustion chamber, 94

Potential energy, 18

pressure, 14

pressure distribution, 52

pressure taps, 151

pressure thrust, 70
products, 85
propellants, 25
PROPEP, 90
pyrolysis, 85
pyrovalve, 102
quadratic Bézier curves, 180
quanta, 86
quenching, 111
quenching distance, 112
range safety, 157
Rankine, 16
ratio of specific heats, 20
reactants, 85
reaction, 12
regression rate, 97
resolved, 15
reversible, 24
Rounding error, 132
run tank, 42
run-valve, 45
safety-factor, 143
saturated, 38
separated flow, 95
shear stress, 144
shifting flow, 91

single-barrier failures, 104
slosh baffles, 133
slosh mass, 41
specific enthalpy of vaporization. *See* specific heat of vaporisation
Specific Gas Constant, 17
Specific heat of fusion, 20
Specific heat of vaporisation, 20
Specific Impulse, 73
splashplate injector, 134
stagnant flow, 23
stagnation temperature, 23
static-firing, 152
stoichiometric, 25
strain, 141
stratification, 46
stress, 141
stress analysis, 141
structural redundancy, 158
subcritical, 36
sub-orbital, 30
subsonic, 64
supercritical, 36
supersonic, 64
Temperature, 15
test stand, 149
the method of characteristics, 70

the method of sect, 142

the minimisation of Gibbs free energy, 89

the rocket equation, 119

thermodynamic properties, 24

thermodynamic system, 21

thermodynamics, 20

Thrust coefficient (C_F), 74

thrust curve, 122

Thrust Optimised Parabolic nozzles, 180

thrust vectoring, 161

tie-lines, 40

total impulse, 122

total temperature, 23

turbulent flow, 96

ullage, 43

under-expanded, 73

Universal gas constant, 17

Urbanski-Colburn valve, 103

valence electrons, 83

vapour pressure, 36

vector, 11

vector diagram, 15

velocity, 54

Von Mises criterion, 143

work, 21

yield, 141

yield strength, 142

yield stress, 142

yielding, 106

Young's modulus, 141

zero-fuel mass, 117

Picture credits

Most of the diagrams were drawn or photographed by the author.

Other images are published with permission to copy, distribute and/or modify under the terms of the GNU Free Documentation License, Version 1.2 or any later version.

The following pictures and diagrams are with kind permission of:

Front cover: The world's largest hybrid engine was tested at NASA's Stennis Space Center August 1999. It was 14 metres long and developed 1,112 kiloNewtons of thrust for 15 seconds: Image by NASA/Stennis Space Center

Page 18: Liquid rocket: NASA

Page 19: Injector plate: Don Young

Page 21: Thrust stand: Aspirespace

Page 26: Fuel grains: Aspirespace

Page 31: 'mount Nitrous': Oxford designers and Illustrators Ltd. courtesy of Rodger Noels

Page 59: Shock diamonds from Aspirespace 'Rickrock' hybrid engine fired at Kingston University, courtesy of Michael Buckley

Page 62: Apollo service module: NASA

Page 75: 'D orbitals' by Wikipedia user Sven

Page 89: 'flame zone' unable to find source of drawing

Page 99: Pump filling: Aspirespace

Page 141: Blockhouse: Aspirespace

Page 142: Test stand: Aspirespace, H20 1st test: Aspirespace

Page 153: Jonathan Rhode's Spirit of Columbia, courtesy of Adrian Hurt